THE MANOR REBORN

THE TRANSFORMATION OF
Avebury Manor

THE MANOR REBORN

THE TRANSFORMATION OF
Avebury Manor

Foreword by
Penelope Keith and Paul Martin

SIÂN EVANS

National Trust

ABOVE: The A over the entrance into one of the walled gardens was put there in the 1980s by Lord Ailesbury: A for Avebury, A for Ailesbury.

PREVIOUS PAGE: Aerial view of Avebury Manor.

First published in the United Kingdom in 2011 by
National Trust Books
10 Southcombe Street
London W14 0RA

An imprint of Anova Books Ltd

Copyright © National Trust Books, 2011
Text © Siân Evans, 2011
Foreword © Penelope Keith, Paul Martin, 2011

By arrangement with the BBC and the National Trust

The BBC logo is a trade mark of the
British Broadcasting Corporation and is used under licence.

BBC logo © BBC, 1996

ISBN: 9781907892240

A CIP catalogue record for this book is available from the British Library.

20 19 18 17 16 15 14 13 12 11
10 9 8 7 6 5 4 3 2 1

Senior Commissioning Editor: Cathy Gosling
Project Manager: Jilly MacLeod
Designer: Rosamund Saunders
Layout Designer: Jacqui Caulton
Copy Editor: Diana Vowles
Production Manager: Laura Brodie
Indexer: Hilary Bird

Colour reproduction by Rival Colour Ltd, UK
Printed in Italy by L.E.G.O. S.p.A.

This book can be ordered direct from the publisher at the website: www.anovabooks.com, or try your local bookshop. Also available at National Trust shops, including www.nationaltrustbooks.co.uk.

CONTENTS

Foreword by Penelope Keith and Paul Martin 6

Foreword

A s a life member of the National Trust, I was surprised and delighted when the BBC approached me in March 2011 with an exciting new project they were proposing in conjunction with the Trust: to transform Avebury Manor and film the process for television. I thought this was an ambitious and groundbreaking project, and so it turned out to be.

Many years ago, while working on a programme on the life and work of Lancelot 'Capability' Brown, I visited and filmed many of the great houses of England, so I thought that Avebury Manor would hold no surprises for me. How wrong I was! The approach through the front garden of the house was idyllic, but on entering it, my impressions altered dramatically. Here was a house that had survived for centuries but required much love and care to ensure its ongoing survival for the enjoyment of visitors in the future. It was clear that a great deal of skill, knowledge, expertise and not a little money would be required to return it to its former glory.

The BBC and the National Trust gathered together a formidable team of experts, craftsmen and volunteers to assist in reclaiming this house for the public. For my part, the making of the television programme has been an eye-opener. I have witnessed the skill and craftsmanship of the silk weavers in Suffolk, the linen weaver in Montrose and the carpetmakers in Craigavon – all a joy to behold.

The BBC and the National Trust have broken new ground at Avebury and all credit is due to them and everyone involved. I urge you all to see and experience Avebury Manor.

Penelope Keith CBE, DL

When the call came telling me about this ambitious project, I couldn't believe my luck. I live just five miles away from Avebury Manor so it's a place I know well and love. Now, through this partnership between the BBC and the National Trust, a joint venture that set out to re-invent the manor as a visitor attraction and to make a television series about it, was an incredible chance to give the place the fanfare I know it deserves. Because the wonder of Avebury Manor has always been overshadowed by the prehistoric stone circle that adjoins it.

It's no exaggeration to say Avebury Manor is a gem of a place ... not the familiar grandiose statement of a 'stately home' but a higgledy-piggledy pudding of a vernacular house – a biscuit-tin cottage with ambitions but retaining an intimacy and a charm that larger houses don't have. Because the real joy of this place is we can all imagine living there. And this, I hope, has informed the whole television and 'visitor experience' ambitions of this project – as we worked to re-create the feel of the different people (Avebury is no 'dynastic' house) who lived and breathed within its walls over nearly five centuries.

It's been said that to look at the steeply gabled east front is to feel the spirit of England – and, sentimentality and patriotism aside, I think it's true. Wiltshire is a part of England steeped in ancient history, prehistory. Most 'intellectual' architecture here feels out of place alongside the vernacular of stone circles and ancient burial sites. And yet Avebury manages to be man-made, even modern in terms of Wiltshire – as natural a marriage with the landscape as it's possible to be.

I hope you enjoy this book which complements and embroiders the TV series. And I hope you visit the manor itself too, to judge if we succeeded – or not – in recreating the feel of a country house through the ages.

Paul Martin

INTRODUCTION

*"… I wanted a home for my family, and I loved
Avebury Manor: and so at a time when most sensible
people are moving out of places like Avebury I moved
in, for I did not feel that a place like this should die.
I feel that the Dunches, Mervyns, Stawells, Holfords
and thousands of others who have known Avebury
would share this view …"*

SIR FRANCIS KNOWLES, BART,
WHO PURCHASED AVEBURY MANOR IN 1955

THE CHALLENGE

I n the heart of rural Wiltshire stands a beautiful old stone house, the tangible evidence of all the hopes and aspirations, ambitious plans and consuming passions of people now long-gone and largely forgotten. Avebury Manor, owned by at least 13 different families or individuals in its 450-year history, is not grand or glorious, but its mellow grey walls, mullioned windows and idiosyncratic gables are brimming with character, and it deserves wider appreciation.

The venerable building has been owned by the National Trust since 1991, when it was bought from the Official Receivers following the owner's bankruptcy. It had long since lost any indigenous contents, so the pragmatic decision was taken to let it out to tenants who would furnish it and live there, with public access allowed on a limited number of days. When the tenancy ended in 2009, the charity wanted to open the house

BELOW: Avebury Manor sits in the heart of rural Wiltshire, beside the famous Avebury stone circle. It dates back to Tudor times, and has been much altered over the years.

more fully to the public. As it was once again empty and almost devoid of contents, National Trust Museum Curator Dr Ros Cleal devised an imaginative temporary exhibition called Moving Out, celebrating the comings and goings of the manor's best-known inhabitants.

In 2010, a casual 'what if?' chat between Basil Comely of the BBC and Alison Dalby of the National Trust resulted in an intriguing idea. The BBC was looking for an empty historic house which could be investigated, decorated and furnished for an innovative TV programme; the plan was to initiate the process, conduct the research using specialist advisers, commission the design team to create a scheme that took the history of the house as its inspiration, and record the transformation as it progressed. The resulting TV series would engage and inform viewers about the practicalities and pitfalls of bringing a historic property back to life. The timing for this unique partnership was perfect; the Trust was interested in 'exploring new ways to give houses more diversity of purpose beyond the traditional approach', in the words of its chairman, Sir Simon Jenkins.

ABOVE: The ancient iron knocker on the front door seems to encapsulate all of Avebury's long history.

ALL EYES ON AVEBURY

A few largely empty houses were possible candidates for this innovative approach; Barrington Court in Somerset and Seaton Delaval, Northumberland, were considered, but Avebury Manor was selected because of its proven popularity with visitors, having attracted 18,000 visitors in 2010. Its domestic scale was also a favourable factor; Kate Shiers, Series Producer, wanted 'somewhere where people could imagine living'.

Basil Comely, the BBC's Executive Producer, summarised it succinctly: 'This is not, strictly speaking, a conservation project, but neither is it just cosmetic. What we want to do is to bring new life to this wonderful old building, to transform key parts of it so that visitors can imagine the changing nature of family life here, on this one site, comparing the Tudor, Georgian, Victorian and Modern eras.'

The funding and leadership for the Avebury Manor project was provided by the BBC, which brought in a team of researchers, innovative designer Russell Sage and expert advisers Dan Cruickshank and Dr Anna Whitelock to create new rooms, imaginatively informed by their historical findings. Collaboration and co-operation with the National Trust's curators, conservators and managerial staff on site was crucial, as a project of this scale needed the co-operation and goodwill of all involved.

The key to success, however, lay in the 20 or more commercial firms, craftspeople, artists and restorers throughout the heritage and historical interior design 'industry' whom Russell approached to work on the project. Without their support and generosity of spirit, and their willingness to dramatically reduce their costs even in a time of recession, the transformation of Avebury would not have been possible. Additionally, the band of local volunteers who willingly gave up their time to transform the kitchen garden and paint the stairs and hallway also contributed greatly to the success of the venture.

THE PLAN

RIGHT: A view through one of the pedimented doorways in the Dining Room shows the studded front door and its elaborate door knocker.

BELOW: Volunteer Jill Lovett gives up her spare time to help decorate the hallway between the Dining Room and the Billiard Room.

In essence, the plan was to design, decorate and furnish nine rooms and a garden in ways that were historically accurate to represent the fabric of daily life during four distinct periods and, through the decorative and furnishing schemes, to tell the story of key people and events in the history of the house. It was a type of project that had never been tried before and to complete it on schedule, in the space of just one summer, required everyone to commit to a massive undertaking.

First, there was a thorough investigation into the lives of the many former inhabitants of the house, involving primary source research into family records, portraits, transatlantic letters, hand-written inventories and wills, and the tracing of objects and artworks known to have been at Avebury Manor. BBC researchers worked with the National Trust's curatorial staff to consolidate agreed knowledge, and the experts, Dan Cruickshank and Anna Whitelock, contributed their profound knowledge of architectural and design history and of social history respectively to help everyone understand the context for each room. They also consulted historic examples, seeking inspiration from properties of an appropriate age such as Dyrham Park in Gloucestershire.

ABOVE: During his presentation to the assembled BBC and National Trust experts, Russell filled the Billiard Room with artefacts and fabrics to give a flavour of the proposed transformation.

Formal scientific research was a necessity; the National Trust commissioned Wessex Archaeology to conduct a thorough investigation into the fabric of the building to identify the various ages of the structure. Dendrochronology (tree-ring dating) was applied to timbers all over the building and provided new information about the age of supporting timbers. Paint analyses were also conducted to see how the house had been decorated in previous centuries.

Russell Sage used the background material and the experts' advice to create each room, using his own knowledge of art, design and historic artefacts. Furniture and fittings were to be newly made, put together from salvaged materials, or bought at auctions, antique shops and markets across the country: all of them had to be durable. The design brief was to create robust furniture that people could use, items that they could touch.

The majority of the new items would look as though they were freshly made, because that is how their original owners would have seen them, and there would be no 'Do Not Touch' signs, no barriers.

The National Trust insisted that everything had to be non-destructive and reversible, so as not to impair the fabric of this Grade I listed building. For their part, the BBC had a modest refurbishment budget; this was a tale in keeping with 'austerity Britain', where pragmatic, affordable solutions were applied creatively and imaginatively. Crucially, where no physical evidence survived or was discernible at Avebury, historical precedents from other houses of a similar vintage could be adopted, if approved by the BBC's experts, the National Trust, and the local authority's Conservation Officers.

BELOW: Avebury Manor is a Grade I listed building and, although it was in need of some loving care, it was important that all major changes were non-destructive and reversible.

DECISIONS AND DEBATE

Some elements of authenticity had to be overruled; the National Trust had to refuse the request for naked flames such as candles and rushlights on the grounds of fire risk, as well as painted panelling, which might damage the wainscoting. But, in discussions with Dan and Anna, Sarah Staniforth, Museums and Collections Director at the National Trust, was able to sanction the decorative painting of the ceiling and frieze in the Tudor Bedchamber; there were colourful painted ceilings in Elizabethan houses like Plas Mawr in north Wales, so it was deemed acceptable at Avebury, so long as it was made apparent to visitors that this particular ceiling had only ever been white.

Russell's design presentation to the assembled BBC and National Trust experts was a *tour de force*; he was, unusually, quite nervous. 'I do this kind of thing all the time for other clients, but this is different and I *have* to get it right,' he said. The subsequent debate was enthusiastic and passionate, and generated a sense of excitement. There were still many obstacles to overcome, from the provision of further design proposals to obtaining all the necessary permissions and commissioning the craftspeople; meanwhile, there was a looming deadline. But as Basil Comely put it, 'We have to solve each problem as it arises. And we cannot eradicate risk, or the whole project would be dishonest from the outset. On the contrary, we have to *embrace* risk and hope that the energy of everyone involved can get us over the hurdles.'

Meet the Team

Penelope Keith

Penelope Keith, CBE, DL, is the much-loved actress inevitably associated with her on-screen persona Audrey fforbes-Hamilton, the upper-class widow in the BBC comedy series *To the Manor Born*. A genuine enthusiast of historic houses and a great supporter of the National Trust, she and her husband Rodney live in a seventeenth-century house in Surrey, and she is fascinated by the idea of all the different people who have inhabited Avebury Manor throughout the centuries. Penelope's role as co-presenter in this series was to explore the house on behalf of the viewer and ask the people involved what could be done to infuse these empty, atmospheric rooms with the personalities and artefacts of bygone ages.

She brought humour and a sense of wonder to this project; on her first exploration of the house, she spotted a hand-written conservators' label hanging from an old bell pull in the Billiard Room. 'It says, "I'm old and badly damaged; don't touch me!"' she announced with a giggle, a pure *Alice in Wonderland* moment.

Paul Martin

Author and TV presenter Paul Martin is fascinated by quirky buildings and their more unusual contents. He was once an antiques dealer with a stall on Portobello Road and also used to run an antiques business in Marlborough, specializing in seventeenth- and eighteenth-century oak furniture. He knows Avebury well, as he and his family live in nearby Seend.

Exploring the house by torchlight with Penelope Keith, Paul commented, 'It's like a rollercoaster ride – some rooms do it for me, and some don't.' Enthusiastic about the restoration project, Paul chaired a public meeting with local people in the library at Avebury Manor to present the concept and to address any concerns before work got under way. He was able to reassure the audience that this would not turn the Manor into an over-commercialised Elizabethan theme park and to address reasonable concerns about traffic congestion and parking. After the design presentation, Paul said, 'This is exciting, I can see this house

coming alive.' He was particularly motivated by the opportunity to visit a number of the specialist craftspeople and to see them enacting the same processes and skills that would have been employed in former centuries.

ABOVE: Presenters Penelope Keith and Paul Martin are both fascinated by Avebury Manor and the people who once lived there.

Russell Sage

Designer Russell Sage is greatly in demand for his innovative, evocative interiors for luxury hotels, fashionable restaurants and private clients, but his passion for old buildings and their contents has always informed all aspects of his diverse design work. When the BBC first suggested this project, Russell leapt at the chance to create a sequence of discrete interiors in an otherwise empty historic house. He wanted to encapsulate the stories of the real people who had lived here at different times, with each of his room schemes indicating the power, the status and the personalities of the owners, as well as saying something about the times in which they lived.

ABOVE: Designer Russell Sage (centre) is flanked by experts Dan Cruickshank and Anna Whitelock. Together they helped to unravel Avebury's past and decide how best to transform each room.

Russell was committed to making each room as evocative as possible, which is why he found it fascinating to see the house through Dan's and Anna's eyes and to understand domestic life in previous eras. To complete his design he returned to historically accurate first principles wherever possible, commissioning authentic pieces from new or salvaged materials, such as accurate carved wooden details, hand-painted Chinese wallpaper and woven rush matting. Every piece needed to be robust, replaceable and capable of withstanding wear and tear from visitors. Velvet ropes were to be banned; Russell's ambition for Avebury Manor was to engage the imagination so that the typical visitor would think, 'So this is what it was like to live here then.'

Dan Cruickshank

Helping Russell and the rest of the team understand Avebury Manor's complex past was Dan Cruickshank, renowned architectural historian, author, journalist, broadcaster and campaigning champion for historic buildings. He is fascinated by this house, not least because there are many unanswered questions about its evolution. 'It's the perfect place to explore the history of the English home, because so many periods and domestic fashions and aspirations are expressed within its architecture,' he said, 'but little in Avebury is quite what it seems. It is a wonderfully evocative and complex building, with portions dating from the mid-sixteenth to the mid-twentieth centuries. Part of the very tangible magic of the manor is its mystery. The richness of the manor's interior and the fascinating array of characters that it has housed and who have left their marks on its fabric offers a fantastic opportunity to re-create and understand the past.

'Indeed, in a sense, almost all the history of the last 450 years of English domestic architecture and design lies latent within Avebury Manor. The object of the current project has been to realise this potential – to take each major room, to build upon its essential architecture and character and to furnish and decorate it in an authentic manner. The result is an evocation of the story of the house and of design in England that reveals the lives of the people who created Avebury and its gardens over the centuries and who collectively made it one of the most enchanting and beautiful houses in Britain.'

BELOW: The graffiti, etched long ago into the stonework in Avebury's south porch, bears witness to the manor's past.

Anna Whitelock

Dr Anna Whitelock is an eminent historian and author and a lecturer in history at Royal Holloway, University of London; *The Manor Reborn* is her first major television role. She worked with Dan Cruickshank to brief Russell and the team on the design schemes and played a key role in advising on historic interpretation and the visitor experience. She led the research on the characters showcased in the house, being particularly keen to portray the motivations and personalities of the various inhabitants, from the Tudor self-made man William Dunch to the marmalade magnate Alexander Keiller.

Anna is a regular media commentator on issues related to the Tudors and Stuarts, the English monarchy, bodies and beds, and on topics related to public history and heritage. She was keen to be involved with the Avebury Manor project because it represented an opportunity to explore on television some of the current challenges facing heritage in Britain. A passionate proponent of public engagement in history, Anna is director of the Centre for Public History, Heritage and Engagement with the Past at Royal Holloway, University of London.

David Howard

David Howard, formerly Head Gardener to HRH Prince Charles, the Prince of Wales, at Highgrove for more than a decade, transformed an abandoned walled garden at Avebury into the type of kitchen garden typical of country estates in the 1880s and 1890s. He designed the garden along traditional lines to historically accurate proportions, with a mixture of ornamental and unusual vegetables, fruit trees including nectarines, pears and apples, cut flowers for the house, glasshouses for delicate plants, and cold frames for cucumbers. David is passionate about gardening organically and planted specifically to encourage bees to visit, as well as providing facilities for the composting of all organic material. He had to work within a very challenging budget, and he had to start at the wrong time of year; ideally he would have wanted a twelve-month preparation period to prepare the ground.

David recruited a devoted band of volunteers consisting of local people and students to undertake the digging, clearing, planting and structural work. As David explained, 'With fifty people doing two hours each, that's a hundred hours, or as much as I could manage to do on my own in three weeks!' He encouraged and inspired them to achieve more than they might have thought possible, and certainly the kitchen garden could not have been transformed into the bucolic idyll it now is without the hard graft of those dedicated volunteers. One last-minute hitch was the discovery that a pair of great crested newts, a protected species, had set up home in the newly erected glasshouse, delaying the laying of the floor. David created alternative accommodation for the reptilian squatters elsewhere in the kitchen garden, and visitors will be able to see Newt Manor for themselves, near the compost bins.

LEFT: Gardener David Howard has designed a kitchen garden typical of country estates in the 1880s and 90s.

The National Trust Team

The National Trust had a core team to oversee all aspects of the transformation of Avebury Manor. Sarah Staniforth, Museums and Collections Director, summarised the charity's view: 'An empty house is like a blank canvas, so this was an exciting opportunity to interpret the interiors of the house in an authentic but imaginative way. Renovation projects are always collaborative, but the Trust had never before handed over a property to another body to reinterpret in a robust, hands-on way.'

Lucy Armstrong, the Project Curator, had previous experience in using physical and archival evidence to inform curatorial and design decisions about interpreting historic houses. 'I was intrigued by the idea of showing rooms of different periods in the house, because many National Trust properties don't date from a single era; they always have a number of complex layers which need to be explained to the visitor,' she said.

Richard Watson, the Project Manager, dealt with the logistics of access on site. 'It was a real privilege to work with the experts involved in the project. I have met exponents of all kinds of disciplines, I've seen how they tackle their particular roles, and I've learnt a lot about the Manor.'

Museum Curator Dr Ros Cleal suggested excellent resources to the BBC's research team and was delighted that the project revealed previously unknown information about the house and its former inhabitants. 'We understand the house and its history so much better than we did and are so glad that we can share all of that with our visitors in such a novel and exciting way.'

For her colleague Dr Nick Snashall, Archaeologist for the Stonehenge and Avebury World Heritage Site, the whole experience was absorbing. 'Much of my role was behind the scenes – dealing with planning consents, arranging archaeological watching briefs when underground work was needed and project managing the Historic Buildings Survey that the Trust commissioned from Wessex Archaeology,' she said. 'For me, that survey is the hidden gem in this story; it has told us things we just couldn't have discovered in any other way.'

Jan Tomlin is the General Manager for the Wiltshire Landscape portfolio. 'From the outset, we were excited about the possibilities offered by this partnership. We shared a common vision of a house brought back to life, where everything had historical integrity, yet was vibrant, colourful and touchable. I feel that the Manor has been brought back to life for another memorable chapter in its history.'

RIGHT: The National Trust team that worked on the project included (from left to right) Ros Cleal, Lucy Armstrong, Richard Watson, Sarah Staniforth and Jan Tomlin.

CHAPTER I

ONE HOUSE...
ALL OF HISTORY

More than once I have taken visitors from overseas to Avebury. Walking from the church to the little museum I have stopped them where, without intrusion, one has a view of the Manor; and I have said, "Do please let this one scene fasten upon your memory. Whatever memories of England you carry away, let this be uppermost. For this, to me, is England."

R.C. HUTCHINSON, NOVELIST (1907–75)

The House Through History

A quintessentially British manor house with a venerable history, Avebury Manor has evolved over centuries, putting out extensions, blending with its surroundings. It is built of grey stone, pierced by windows of leaded panes and surmounted by a medley of gables, pitched roofs and chimneys, added at different times over the last 450 years.

It is not a great house, massive in scale, occupied by powerful figures at the vortex of political change, nor is it an architectural gem, showcasing the work of a single great visionary. No worldly and fashionable man of affairs had a hand in its creation and there is no array of the glittering prizes of a former age. On the contrary, it is a relatively modest home in which successive generations of families have lived out their lives in varying degrees of comfort and happiness.

BELOW: The east front of Avebury Manor is a delightful patchwork of gables, chimneys and mullioned windows, dating from the sixteenth and early seventeenth centuries.

The changing fortunes of the house are not immediately apparent from its tranquil atmosphere and the sense of timelessness. Avebury Manor was not immune to the turbulent events of history: two of its former owners were arrested by the Crown, one of whom was imprisoned in the Tower. Myths also abound – tradition has it that the house grew up around the Great Hall of a Benedictine priory and that peripatetic royalty slept here, but there is no evidence to support either story.

What is certain is that the manor served as a farmhouse for nearly 100 years before being

Avebury Manor was not immune to the turbulent events of history: two of its former owners were arrested by the Crown, one of whom was imprisoned in the Tower.

restored to its former glory by owners who venerated the past, and has been owned variously by a well-connected civil servant, a serving soldier, a lawyer, a scientist, and a millionaire marmalade magnate.

While there is no evidence that the manor house incorporates a medieval structure, it is known to lie within the vicinity of a much older building. The recorded history of the site dates back to the early twelfth century when the Abbey of St Georges de Boscherville in Normandy decided to establish a Benedictine priory at Avebury, inhabited by a small group of monks who employed local labour to run the large estate. This was prime agricultural land and the priory prospered; according to a document of 1324–5, the Abbey owned 600 sheep and the priory contained a kitchen, bakehouse, brewery, cellar and dairy, as well as a hall.

ABOVE: An attractive lavender-lined pathway leads up to the original Tudor entrance, above which sits an Italianate plaque, placed there in the twentieth century.

ABOVE: With its stone-tiled roof and rugged sarsen walls, the twentieth-century library (on the left) blends seamlessly with older parts of the building.

The evolving house

The manor house lies close to the venerable church of St James and adjacent to the Avebury stone circle, which so intrigued John Aubrey, seventeenth-century antiquarian and man of letters. It is approached by a classical gateway which visitors reach by walking across what was once the estate farmyard, with its thatched great barn and picturesque sixteenth-century dovecote. Ahead stands the east front of the original house, built around 1557 by William Dunch and subsequently extended by Sir James Mervyn. The façade is a delightful patchwork of gables and mullioned windows, topped with chimneys of various ages.

The south wing of the manor dates from about 1600, when Mervyn built a new wing at right angles to the older house, comprising a Great Hall with a Great Chamber above, and a new main entrance marked by a fine classical porch. At the same time, if not earlier, the southern end of the original house was extended to provide a new parlour and bedchamber, both with exceptional plasterwork ceilings. The south frontage you see today, however, is the result of extensive remodelling carried out in the mid-eighteenth century, when the gables were replaced by a parapet and the interior rooms were upgraded. The original Great Hall was transformed into a dining room and the chamber above into a State Bedroom with a magnificent coved ceiling.

At the western side of the house stands a library which, although it was constructed in the 1920s, is convincingly harmonious with the architectural style and materials of the Tudor structure; a flight of steps links the library with the formal garden.

JOHN AUBREY

John Aubrey (1626–97) was an antiquarian and man of letters whose collection of gossipy and controversial biographies known as *Brief Lives* became a hit one-man show in the twentieth century. A Wiltshire man, he discovered the stones of Avebury while out hunting in 1649 and later described the prehistoric site as 'the most eminent and most entire monument of this kind in the Ilse [sic] of Great Britain . . . I was wonderfully surprized at the sight of those vast stones of which I had never heard before.'

John Aubrey's written descriptions, and a comprehensive plan of the Avebury Circle, the Avenue and the Sanctuary, were featured in his *Monumenta Britannica* (compiled *c.*1665–93 but never published), on instructions from Charles II. The plan is surprisingly accurate given the limitations of surveying at the time and the ignorance of ancient indigenous cultures.

BELOW: The south wing originally dates from about 1600, but was substantially remodelled in the mid-eighteenth century, in keeping with contemporary taste.

The History
of Avebury

1114
Benedictine priory founded
on the site

1378
Priory is dissolved

1411
Priory passes into hands of
Fotheringhay College

1545
Fotheringhay College gives up
ownership of the estate

1547
William Sharington acquires the
Avebury estate

1551
William Dunch, Auditor of the
Royal Mint, buys Avebury for £2000
and builds the manor house c.1557

1595
Dunch's widowed daughter-in-law,
Debora, and her second husband,
Sir James Mervyn, become
owners of Avebury outright

1640
Sir John Stawell buys Avebury for
£8,500 from Debora's son, William
Dunch; his son then grandson inherit

30

The Inhabitants
of Avebury

Avebury has passed through many hands in its long history. Due to wars with France, and the decimation of the population through successive waves of the Black Death, the Benedictine priory that once stood on the site was dissolved in 1378. The Crown confiscated the estate, which was subsequently run by a succession of favoured royal servants. In 1411 the priory passed into the ownership of Fotheringhay College for chantry priests in Northamptonshire, which received the revenues until it gave up the estate in about 1545.

Avebury was sold in 1547 to William Sharington of nearby Lacock, one of the many ambitious men to have profited under Henry VIII, and who were now keen to establish themselves as landed gentry, like the old nobility. For a long time it was thought that Sharington may have rebuilt Avebury Manor, incorporating parts of the old priory into its structure, but recent evidence suggests that the house was newly built in about 1557. As Sharington sold the estate in 1551 to a courtier named William Dunch, it is now clear that Dunch was responsible for building the new manor house, although it is not known what became of the priory. Dunch was succeeded through marriage by the Mervyns, who built the south wing and a classical porch bearing their initials.

By the end of the seventeenth century, the house was in the hands of Sir Richard Holford, whose grandson remodelled it to create a dining room and fashionable bedroom on the first floor. Following the tenure of the Holfords, Avebury Manor was then left as a bequest to the wife of Sir Adam Williamson, former Governor of Jamaica, who retired there. After a long and action-packed life in far-flung places, which provided ample opportunity to be killed by a stray bullet, Williamson died following a fall from a chair in the dining room.

RIGHT: The Mervyns reorientated the manor by building a south-facing wing at right angles to the original block. They marked their new main entrance with a splendid classical porch, complete with fluted pilasters and a plaque bearing their initials and the date 1601.

1694
Sir Richard Holford, Master in
Chancery, buys Avebury for £7,500;
a succession of relatives inherits

1789
Ann Williamson, wife of
Lieutenant-General Sir Adam
Williamson, inherits the manor
from Arthur Jones

1798
Sir Adam Williamson dies at Avebury
and house passes to Richard Jones,
Arthur Jones's nephew

1816
The Kemm family move in as
tenant farmers

1873
Brewer and politician Sir Henry Meux
buys Avebury; son Henry inherits
upon his death in 1883; the Kemms
remain as tenants until 1902

1900
Sir Henry Meux junior dies

1902
Meux's widow lets the estate to
Colonel Leopold and Nora Jenner

1907
The Jenners buy the estate

1929
The Jenners let the manor to the
Benson family

The History
of Avebury

ABOVE: This old photograph from 1899 shows the intimate relationship between the farm buildings and the ancient standing stones during the Kemms' tenancy of Avebury. Several buildings were subsequently torn down to aid Keiller's excavation of the site.

1937

Alexander Keiller buys the house from the Jenners and establishes the Morven Institute of Archaeological Research

1943

Keiller sells most of the property, including the stone circle, to the National Trust for £12,000; the Trust declines to buy the house as well

1955

Sir Francis Knowles buys Avebury; Keiller dies and his widow later donates his collections to the nation

1976

Sir Francis's widow sells Avebury Manor to Michael Brudenell-Bruce, 8th Marquess of Ailesbury

1981

Mr and Mrs Nevill-Glidden purchase Avebury Manor

1988

Kenneth King buys the manor house for about £1 million and plans to turn it into an 'Elizabethan Experience', causing much local controversy

1991

The National Trust buys the house from the Official Receivers following King's bankruptcy

A family of Wiltshire tenant farmers called the Kemms occupied the house for much of the nineteenth century, during which time – despite minor alterations – benign neglect took over. In 1902, however, the manor was leased to Lieutenant-Colonel Leopold and Mrs Jenner, whose pioneering passion for old buildings was boundless. After living there for five years, the couple bought Avebury Manor and lovingly restored it in the vernacular style, trawling outbuildings, farm sales and auction houses to acquire authentic architectural salvage such as oak panelling which they could use in the interiors. They built the new library on the west side of the house and also laid out a highly attractive and structured garden.

The Jenners were forced to relinquish Avebury in 1929, possibly as a result of financial difficulties. It needed a millionaire with vision to take it on, and fortunately Alexander Keiller was just the man. In 1937 he made the manor house his home, and the base from which he organised annual excavations of the vast prehistoric site on his doorstep in the years before the Second World War. The house accommodated

Keiller's archaeological foundation, but eventually it was sold privately to a succession of owners. In the late 1980s an entrepreneur called Ken King had a much publicised plan to turn the house and grounds into a Tudor theme park, which caused a great deal of local controversy before King succumbed to bankruptcy. In 1991 Avebury Manor was bought by the National Trust, and, as it was virtually empty of furniture and fittings by this time, it was leased to a private tenant with an arrangement that it should be open to the public for a limited time only. Since its transformation in 2011, however, the aim is to keep it open on a regular basis during the visitor season.

The Kemm family occupied the house for much of the nineteenth century, during which time benign neglect took over

BELOW: In 1938, the seventeenth-century stable block to the east of the manor house became home to the Alexander Keiller Museum. Here, finds from Keiller's excavations are still on display, including prehistoric tools, reconstructed pots and ancient skeletons.

TRACING AVEBURY MANOR'S PAST

In 2011 the National Trust commissioned a full historic building survey on Avebury Manor. Previous investigations had been conducted in the early 1990s, but it was still not known how this house had evolved and, critically, when different elements of it had been built. Scientific analytical methods had progressed so much in the intervening two decades that a new survey, conducted by Wessex Archaeology Limited, was timely.

Avebury Manor did not give up its secrets lightly; previous owners had repaired old damage and introduced salvaged antique panelling, while primary documentary sources, such as contemporary letters and accounts, were unusually scarce. Wessex Archaeology scoured primary and secondary research material, turning up crucial new evidence. Their

BELOW: Thanks to the recent archaeological survey, it has been possible to piece together a likely development for Avebury Manor over the centuries. The first plan is reliably based on a sketch of the Manor dating from 1695 (see page 36) and shows the original house from c.1557 with later extensions, most of which were built by the Mervyns in c.1600. The remodelling in c.1740 made little impact on the ground plan; the last major extension was in the early twentieth century, when the Jenners built a new library.

16TH AND 17TH CENTURIES

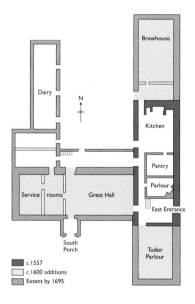

18TH CENTURY

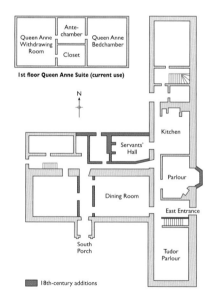

findings, including a full analysis of the fabric of the building, were collated in a detailed survey, and revealed that the initial house had been built on a modest scale around 1557; this had then been extended to the north, south and west around 1600, and remodelled in about 1740.

Vital to determining the age of the building was the science of dendrochronology, or tree-ring dating. Every year, trees in various parts of the country grow at different rates, according to climatic conditions, and these unique variations have been 'mapped' to span many centuries. Core samples taken from timbers in buildings can therefore reveal when a tree was felled, helping to pinpoint the construction date. At Avebury Manor, scientists were even able to identify beams in different parts of the roof that have been taken from the same oak tree. Dr Nick Snashall commented, 'Wessex Archaeology used everything from laser scanners to good old pen and ink to unlock the story of the manor. The most exciting moment was watching Robert Howard, the tree-ring specialist, taking samples and knowing that the tiny piece of wood he held in his hand would totally transform our ideas about how the house was built."

ABOVE: With the aid of a special drilling device, wood cores were removed from timbers and labelled. Analysis of the tree rings enabled scientists to date the timber.

19TH CENTURY

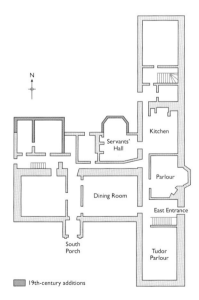

20TH CENTURY

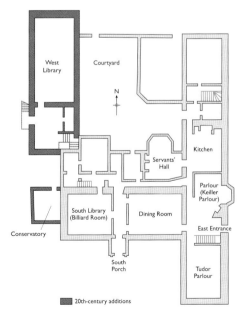

AVEBURY'S GROUNDS AND SURROUNDS

Avebury Manor has long been a working estate; for centuries, successive owners have acquired, sold, leased out and farmed the rich agricultural land around it and provided employment for local people as labourers, farmhands, carpenters or servants. Many farm outbuildings were added by Baron Stawell in the late seventeenth century; a sketch map of the Manor and garden drawn in 1695 refers to the 'New Barne', the 'Old Barne', 'Pidgeon House', granary, coach-house stables and 'Horse Pool'.

As well as self-sufficiency in terms of meat and grain, manor houses prided themselves on growing fruit and vegetables for the household's consumption. At Avebury Manor a walled

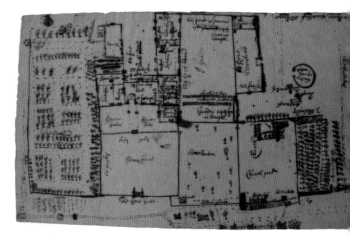

ABOVE: A pen and ink sketch from 1695 shows the manor house (towards the top, left of centre) and the surrounding gardens as they would have looked during Holford's residency.

BELOW: Summer planting thrives in the box-bordered beds that make up the enclosed garden to the south-east of the manor.

kitchen garden was created by the Jenners in the early twentieth century, but there may have been a kitchen garden here in earlier times, tended with care by a head gardener and his assistants.

The grounds of Avebury Manor

The grounds of Avebury Manor are a delightful, meandering complex of stone walls, interlinked enclosed gardens and sculpted topiary. There are elderly vines trained up sunny brick walls, and shaggy drifts of fragrant honeysuckle, cotoneaster, blue pansies and borage, alive with honeybees. Footpaths made of ancient and worn flagstones crisscross the lawns and provide avenues between pairs of massive sculpted yews that seem to mimic the world-famous standing stones just beyond the garden boundaries.

The gardens as they now appear were created by the Jenner family as a sequence of linked walled gardens, their shapes designed to follow the remains of old hedges and walls. Some of the walls are believed to date back to the early

ABOVE: The dramatically sculpted box topiary in the walled garden that lies to the south-west of the manor was created by the Jenners in the early years of the twentieth century.

AVEBURY PET CEMETERY

One poignant feature in the garden is the pet cemetery with eight gravestones, the most recent of which commemorates 'Henry. A Special Friend, much missed. 25 April 1995 to 11 June 2004.' Among the older headstones is one engraved 'Darling Phaeton. Our Joy for 11 years. Born Jan 12 1923. Died Jan 26 1934', and there is an 'Adorable Daisy' too. But the earliest stone dates from the 1850s; it is oddly touching to think that the tradition of interring beloved pets here has lasted for 150 years, despite the many times the house has changed ownership.

TOPIARY

The art of training and clipping trees and shrubs to form sculpted, ornamental shapes dates back thousands of years to ancient Egypt and Persia. The Romans were enthusiastic practitioners, engaging landscape gardeners – the *topiarius* – to create elaborate designs including, according to Pliny the Elder, hunting scenes and fleets of ships. With the decline of the Roman Empire the practice virtually died out, but a resurgence of interest in all things antique during the Renaissance saw it come back into favour.

Topiary reached a peak of popularity in Britain in the seventeenth century following Charles II's return from exile in France, where the court had much admired the formal gardens of Louis XIV's landscape gardener, André Le Nôtre. But the eighteenth-century vogue for the Picturesque and for landscape gardening as practised by Lancelot 'Capability' Brown resulted in topiary's fall from grace and the widespread destruction of formal gardens. It was the Victorians, with their enthusiasm for revival styles, who made it fashionable once more. By the late 1800s, under the influence of the Arts and Crafts Movement, gardeners such as Gertrude Jekyll were designing gardens as a series of outdoor rooms in which sensuous colours and textures were combined with traditional garden crafts such as topiary.

ABOVE: Punctuated by rising topiary cones, the elegant parterre at Ham House in Surrey is based on a design from 1671.

BELOW: Hidcote Manor, Gloucestershire, is famed for its Arts and Crafts garden 'rooms'. In the Pillar Garden, yew topiary towers over beds of allium and peony.

eighteenth century, such as the long curved wall to the west of the manor. Hard-edged structural elements are softened by large-scale topiary in box and yew. Outside the library the topiary garden contains a rectangular pool, flanked by geometric hedges whose intricate design recalls one of the plaster ceilings inside the house. Elsewhere, there are colourful annual beds edged with box, and an orchard surrounded by well-coiffed yew.

Each of the walled gardens is very warm, being protected from the wind, and there are fruit trees espaliered against sunny west-facing brick walls. The gardens are notable for the use of fragrant plants; the main path leading to the house is lined with lavender bushes, while elsewhere there are roses, thyme and jasmine.

BELOW: The geometric box topiary created by the Jenners to the west of the library was based on the design of the plaster ceiling in the Tudor Parlour.

ABOVE: The long curved wall that encloses the south-west garden was, according to tradition, built by Sir Richard Holford. It was incorporated into the new garden scheme by the Jenners.

Avebury village

The London to Bath road created by the Romans ran past Silbury Hill, the man-made chalk mound that is part of the complex of Neolithic monuments around Avebury, and archaeological evidence of a Roman village of considerable size has recently been found at its base. However, after the Romans retreated from Britain in the fifth century CE an Anglo-Saxon colony was set up at Avebury, less than 1.6km (1 mile) to the north and close to Winterbourne Stream. This is evidenced by two huts with sunken floors excavated in the 1970s and the remains of other structures found in the northern side of the modern visitor car-park. A larger village gradually evolved at Avebury, with wooden houses protected from marauders by ditched enclosures, and it is possible the whole village was defended by a surrounding bank and ditch.

Today's Church of St James retains visible elements of the Anglo-Saxon structure built over 1,000 years ago

With the coming of Christianity, a church was erected in Avebury. Today's Church of St James retains visible elements of the Anglo-Saxon structure built over 1,000 years ago, which seems to have been a 'minster' church – an establishment held by the Crown and taking the rents from a large portion of land. The entry in the Domesday Book, compiled in 1086, reads: '*Rainboldus presbyter tenet ecclesiam de Avreberie, ad quam pertinent II hidae. Valet XL solidi.*' ('Reinbald the priest holds Avebury Church to which 2 hides belong. Value 40s.') Hides were units of land of about 50ha (120 acres), so Reinbald held about 100ha (240 acres) at Avebury and

BELOW: The village of Avebury sits in the midst of the ancient stone circle. Many houses were constructed using broken-down megaliths; others were built with more traditional materials such as chalk, timber and daub.

at that time the livings from 15 other churches too. After his death during the early twelfth century Cirencester Abbey was endowed with his land, including Avebury Church and the land that went with it.

An archaeological landscape

The chalk landscape around Avebury contains one of the greatest surviving concentrations of Neolithic and Bronze Age monuments in Western Europe. As well as Silbury Hill, the largest prehistoric man-made mound in Europe, Avebury lies close to one of the longest burial mounds in Britain, West Kennet Long Barrow, and the longest avenue of standing stones in Britain, West Kennet Avenue. To the north, 1.6km (1 mile) away, lies Windmill Hill, one of the largest settlement sites of the earlier Neolithic period.

The complex also includes West Kennet, comprising the remains of two palisaded enclosures, and a monument known as the Sanctuary, once the site of concentric rings of stones and wooden posts. Dotted all around are circular burial mounds dating from the Early

Bronze Age. Avebury's most famous monument, however, is the mighty henge, containing the largest stone circle in the world and rivalling Stonehenge 38km (24 miles) to the south in its impact on the landscape.

Today, most of this archaeological site, along with much of the village that lies inside the henge, is owned by the National Trust, and more than 300,000 visitors arrive each year, eager to see one of Britain's greatest monuments.

ABOVE: Dating back to Anglo-Saxon times, St James's church was expanded over several centuries. The elaborately carved main doorway is Norman and the tower is fifteenth century.

LEFT: West Kennet Long Barrow is one of Britain's most accessible Neolithic chambered tombs. Remains show that nearly 50 people were buried here before the chamber was blocked off.

AVEBURY STONE CIRCLE

At the centre of the Avebury complex of Neolithic monuments is a henge – a circular bank and ditch enclosing a vast circle of standing stones – within which lie two smaller stone circles. Built between 3000 and 2000BCE and measuring about 420m (460yd) across – one of the largest henges in Britain – it is a Scheduled Ancient Monument. As one of the most important Neolithic sites in Europe, the whole complex, including all the surrounding monuments, was designated a World Heritage Site in 1986.

Though this part of Wiltshire is mostly chalkland, the area around Avebury contains a geological oddity – enormous boulders made of sarsen (a hard, grey sandstone), probably formed about 30 million years ago. The outer circle of the henge originally contained about 100 of these sarsen boulders, some weighing more than 40 tonnes. There is also an avenue of paired stones leading from the southern entrance to the henge, and the remains of a second avenue leading from the western edge, which it is assumed had great ceremonial or ritual importance to Stone Age and Bronze Age peoples.

The great mystery surrounding Avebury is why it was built. The site may have been a focus for ancestor worship, as human bones found in the ditch might infer, or it could have been associated with celebrating certain times of the year. It could not have been defensive, however, because the great ditch lies inside the henge.

BELOW: The vast circular bank and ditch at Avebury enclose the largest stone circle in the world. This huge monument took an estimated 1.5 million hours to build, using little more than antler picks and rakes.

RIGHT: One of the mighty sarsens in the outer circle towers over the landscape. Many of these megaliths were long gone by the time Keiller excavated the site.

WILLIAM STUKELEY

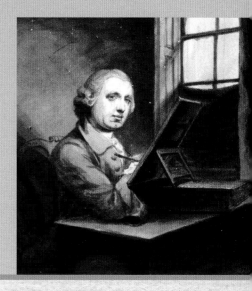

A committed antiquarian, William Stukeley (1687–1765) visited Avebury every year between 1719 and 1724 and drew up detailed, measured plans and drawings, adding Beckhampton Avenue to the monuments previously recorded by John Aubrey. In 1743 Stukeley published *Abury: A Temple of the British Druids*, which added to the mystique surrounding the site. His research was invaluable as it was undertaken at a period when many of the monuments were being destroyed to produce building stone, and his findings were crucial to the rediscovery of the Sanctuary and the Beckhampton Avenue in the twentieth century.

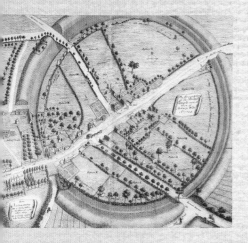

ABOVE: An overhead plan of the henge, published in *Abury: A Temple of the British Druids* (1743). Stukeley's notes, maps and diagrams provided future scholars and archaeologists with an invaluable account of the site for centuries to come.

By the Iron Age, the henge site seems to have fallen into disuse. Following the Roman invasion, Roman settlements were established around Devizes, and there is even a suggestion that Romans and Britons visited Avebury out of curiosity, using the newly made road and discarding broken pottery in the ditches on the site. The enormous pagan monument mystified the invading Anglo-Saxons and was mistrusted by the devout Christians of the Middle Ages, who built St James's church and a priory just outside the circle in an attempt to counteract its baleful influence. Over the years, concerted efforts were made by the people of the village to push over and bury some of the stones; later, many of the stones were broken up using strategically lit fires to create fractures, so that they could be used for local buildings. Concrete markers and a few grassy hollows show where the vanished stones once stood.

Antiquarians at Avebury

Some early antiquarians showed an interest in the site and speculated about it. John Aubrey wrote about it in his unpublished manuscript *Monumenta Britannica*, and William Stukeley left behind useful diagrams showing where the stones would have stood before human interference. He expressed the view that Avebury had been constructed by druids during what we now call the Iron Age, but this theory has been disproved as the complex is at least 2,000 years older. Other competing historical theories included the idea that Avebury and

Stonehenge were both built by Phoenicians, or that the Avebury circle commemorates the death in battle of King Arthur and his nobles; it was even suggested that Native Americans had crossed the Atlantic in order to create a sequence of megaliths across Britain. The truth is that it is not known what the purpose of this huge complex was, but it undoubtedly had great value to the peoples who created it.

In 1873 a Victorian MP, Sir John Lubbock, pushed through a bill to preserve ancient monuments and, upon being elevated to the peerage in 1900, chose the hereditary title Avebury after the ancient druidical site which he had long fought to save from being 'destroyed for the profit of a few pounds'. Harold St George Gray of Avebury then conducted archaeological excavations in the first years of the twentieth century, determining the unexpected size of the huge ditch.

It was in the 1930s, however, that Avebury's final salvation occurred, thanks to the tenacity and dedication of archaeologist Alexander Keiller. Marconi, the radio pioneer, had previously expressed an interest in erecting an aerial on Windmill Hill, so Keiller bought the site to prevent future development. Intrigued by the entire archeaological complex at Avebury, he eventually acquired 384ha (950 acres), along with the manor house, all of which eventually passed to the National Trust.

BELOW: The circular bank and ditch at Avebury are clearly articulated when viewed from the air. Also apparent is the extent to which buildings colonised the henge as the village expanded.

CHAPTER II

AVEBURY IN THE TUDOR AGE

THE TUDOR ERA WAS A DANGEROUS AGE, BUT CLEVER PEOPLE LIKE WILLIAM DUNCH MADE THEIR FORTUNES BY OFFERING UNSWERVING LOYALTY TO THE MONARCH OF THE DAY. SUCH MEN BUILT SOLID HOUSES AT THE HEART OF PROSPEROUS COUNTRY ESTATES AND FILLED THEM WITH FURNITURE, TAPESTRIES, ART AND ARTEFACTS THAT WERE NOVEL AND SOPHISTICATED IN TASTE. THEIR HOMES ENCOMPASSED ALL THE SYMBOLS OF WEALTH AND WERE INTENDED TO CONVEY AN UNMISTAKABLE MESSAGE ABOUT THE OWNER'S SOCIAL STATUS.

LIFE IN TURBULENT TIMES

At all levels, sixteenth-century English society was in a state of flux. The founder of the Tudor Dynasty, Henry VII, had snatched the throne and was determined to use any measures to retain it. Equally driven were his four successors; Henry VIII's desperation to have a legitimate male heir generated consequences that lasted for centuries. The period between the 1520s and 1558 encompassed Henry's many marriages, the break with Rome and the Dissolution of the Monasteries, the monarch's death in 1547, his succession by his short-lived Protestant son Edward and then the 'bloody' reign of his Catholic daughter Mary. This was a particularly dangerous and unsettling time, when plots were suspected and proven on all sides, and men were frequently sent to the Tower on suspicion of treason.

Some people, however, prospered despite the uncertain times and, in the scramble to buy monastic property following the Dissolution of the Monasteries, many people became very wealthy. After the surprisingly

LEFT: Henry VIII, seen here in a painting after Hans Holbein, oversaw a speculation boom as his subjects scrambled to acquire monastic buildings and land.

TUDOR TIMELINE

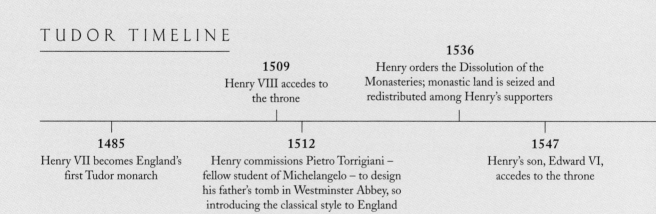

1509
Henry VIII accedes to the throne

1536
Henry orders the Dissolution of the Monasteries; monastic land is seized and redistributed among Henry's supporters

1485
Henry VII becomes England's first Tudor monarch

1512
Henry commissions Pietro Torrigiani – fellow student of Michelangelo – to design his father's tomb in Westminster Abbey, so introducing the classical style to England

1547
Henry's son, Edward VI, accedes to the throne

peaceful accession of Elizabeth I to the throne in 1558, the new nobility benefited from the relative stability of the nation, the general increase of wealth, and the growth of international trade, overseas expansion and global exploration. Even the threat from the Spanish served to unite the country against a common enemy.

Meanwhile, the English Renaissance saw an explosion of secular culture and learning, and an artistic flowering that culminated in the works of William Shakespeare and the building of such magnificent houses as Hardwick Hall in Derbyshire and Montacute House in Somerset. Artistic endeavour was aided by the invention, in the previous century, of the printing press with its movable type. By the 1560s illustrated pattern books and architectural treatises from the Continent were kindling an enthusiasm for Italian and French Renaissance designs, and classically inspired features such as columns and acanthus leaves began to permeate the decorative arts and architecture.

The Reformation and the new landowners

Henry VIII decimated the old order, punishing anyone who dared to stand against him. The schism from Rome enabled him to seize the considerable assets of the Catholic Church in England and Wales – land, buildings, livestock, plate, vineyards, treasures and textiles. The monasteries were closed down, and their lands sold to the wealthy, with the revenue going to the Crown.

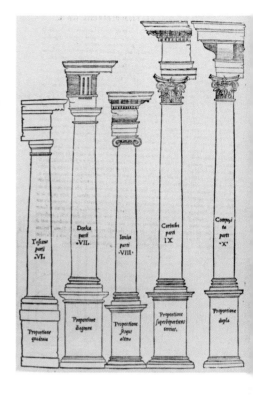

ABOVE: Sebastiano Serlio's *L'Architettura* (published 1537–51) was hugely influential in spreading Italian Renaissance designs throughout northern Europe. This woodcut shows the five orders, or styles, of column used in Roman architecture.

1553	1588	1603
Henry's elder daughter accedes to the throne as Mary I; her persecution of Protestants earns her the nickname 'Bloody Mary'	Spain launches the Armada against England but is defeated by the English fleet and stormy weather	The Tudor era ends with Elizabeth's death; James VI of Scotland, son of Mary, Queen of Scots, becomes James I – England's first Stuart monarch

1558		c.1595
Elizabeth I, Henry's daughter by Anne Boleyn, accedes to the throne; a time of increasing prosperity and exploration begins		Shakespeare writes *Romeo and Juliet*

Elizabeth I

ABOVE: William Sharington acquired Lacock Abbey in Wiltshire, and converted it into a house *c.*1540.

BELOW: Bess of Hardwick married four times and, in doing so, became one of the richest women in England.

Henry rewarded his supporters for their loyalty to the Crown and to the Protestant cause. The favoured 'rising men', the merchants, lawyers and administrators who came from obscure backgrounds and served the monarch, such as William Dunch who bought Avebury Manor, were motivated by the offer of a knighthood, a coat of arms, crests and mottoes, which provided an instant pedigree.

Wealth and status

Conspicuous consumption was actively encouraged, in imitation of the monarch and his court, with more opulent dress for both men and women, the growth of secular paintings including portraiture, and a desire for luxury guarded. Marriage was a business transaction, an investment; many people remarried on the death of a spouse, and their respective children might also be paired together in order to consolidate rather than divide the combined estates. A Tudor wife was expected to oversee the running of the large household and to deputise for her husband in matters of business in his absence. Townswomen were often very active partners in family businesses such as goldsmithing or glovemaking, while their sisters in the country tended to run the house and provide leadership and instruction for all the inhabitants.

THE TUDORS AT HOME

Substantial houses in rural areas were largely self-sufficient, brewing their own beer, grinding the flour, baking the bread, making dairy products, keeping fish-ponds and dovecotes for non-meat days (pigeons and doves being considered not strictly 'meat'), and making medicines, cordials and perfumes in the stillroom. However, a house such as Avebury Manor was not isolated from city life; it was on the main road between London and Bath, and the Dunch family who lived here for decades had important connections at court as well as other properties in Wiltshire and Berkshire. They could write to merchants in London to order goods, or family members might receive a commission to buy goods and dispatch them to Avebury.

Maintaining a large household remained a status symbol throughout the Tudor era, but there was also a sense of *noblesse oblige*, supporting others by providing them with employment. Menservants predominated; women's roles were confined to nursemaiding and caring for children. All inmates, whether blood relatives or servants, apprentices or dependents, were considered to be the master's 'family', literally meaning 'those familiar to me'.

BELOW: Substantial houses needed substantial kitchens. At Buckland Abbey, Devon, the kitchen was resited to be near the monastic Great Hall, and was dominated by two large open hearths used for cooking.

ABOVE: The moat at Lower Brockhampton originally encircled the house and may have acted as a status symbol as well as a defensive barrier. The gatehouse was built some years after the moat.

THE TUDOR HOUSE

In previous centuries, substantial country houses had been fortified to deter intruders; some had a moat and gatehouse, as at Lower Brockhampton in Herefordshire, built in the late fourteenth century. However, with the coming of peace, new buildings could be more open and less defensive. Half-timbering, so typical of English architecture, was the standard building method in wooded areas, where the raw materials were plentiful, or in towns where timber could easily be brought in by boat. At Avebury, set on chalk downs littered with flint and sandstone sarsens, stone was the natural building material for buildings such as the manor house and church.

Master builders and masons were responsible for the design and construction of houses in this era. They relied on pattern books and line drawings to agree the details of a commission with the landowner. Master builders were often recommended by a former client, and of necessity these highly skilled and respected craftsmen travelled around the country to work on site. They would subcontract local men with some experience to erect the building, but once the structure was complete, specialists such as plasterers or master carvers would be brought in to finish the interiors.

MONTACUTE HOUSE

This Elizabethan house near Yeovil, Somerset, was built in the late sixteenth century for a lawyer, Sir Edward Phelips, Speaker of the House of Commons and the prosecutor of Guy Fawkes. A more sophisticated and worldly house than Avebury Manor, Montacute benefited from the same good fortune of being occupied and restored in the early twentieth century by a buildings conservation enthusiast; in 1915 it was rented by Lord Curzon, former Viceroy of India, who had a passion for acquiring and restoring old houses such as Bodiam Castle in Sussex and Kedleston Hall in Derbyshire. Curzon spent a fortune on the largely empty property, asking his mistress, Elinor Glyn, to arrange the removal of 'inappropriate' decorations, buying Tudor era furnishings, and commissioning rush-matting for the floors – at about the same time that Leopold and Nora Jenner were restoring Avebury Manor. Both houses were lovingly cared for by pioneering conservationists, and both ended up in the care of the National Trust. In the case of Monacute, this was thanks to Ernest Cook, the grandson of Thomas Cook, the travel entrepreneur, who saved the property from demolition in 1931 by funding its purchase.

ABOVE: The Great Hall at Montacute is not as grand as its medieval predecessors, as was common by the Elizabethan age.

BELOW: In 1931 Montacute House was valued 'for scrap' at £5,882. Yet, thanks to Ernest Cook's timely intervention, it is now one of England's best preserved Elizabethan mansions.

Windows were essential for letting in as much daylight as possible, but often they were unglazed and instead covered by sturdy, lockable wooden shutters to give privacy and to keep out draughts and intruders. By the Tudor era glass was still an expensive commodity, so having many windows was a status symbol, as at Hardwick Hall in Derbyshire and Montacute House in Somerset. Small panes of glass were held in place by a leaded trellis, and when a family moved house they might well take their windows with them to the new location.

The Great Hall

At the heart of a medieval manor house lay the Great Hall, the focus of domestic life where the family, their employees and any visitors would dine together and socialise, mostly at long refectory tables, seated on benches or stools. The hall was flanked at one end by the owner's private apartment and at the other by the kitchen and other service rooms, which were usually screened off from the main space to form a 'screens passage'. The assembly in the Great Hall was convivial, but social deference was expected; the householder and his family would sit on a raised dais, and use chairs rather than stools to indicate their higher rank. Vita Sackville-West gives a vivid description of life in the Great Hall at Knole, in Kent, in *Knole and the Sackvilles* (1922):

BELOW: Dating from the sixteenth century, the Great Hall at Hardwick Hall in Derbyshire is dominated by the plasterwork overmantel, which bears the Hardwick coat of arms supported by two stags.

'Great state was observed here once, when well over a hundred servants sat down daily to eat at long tables in the Great Hall . . . all coming in from their bothies and outhouses to share in the communal meal with their master, his lady, their children, their guests, and the mob of indoor servants whose avocations ranged from His Lordship's Favourite through innumerable pages, attendants, grooms and yeomen of various chambers'

As prosperity grew, so too did the desire to express privilege through increased privacy; manor houses became more complex, with two or more storeys and a greater number of specialized rooms. By the sixteenth century it was more common for the Great Hall to be floored over to accommodate a Great Chamber on the floor above, as in the south wing at Avebury Manor.

Hearth and home

Until the fourteenth century, most heating was provided by an open hearth in the centre of the Great Hall. The fire generated heat, some light and often a great deal of smoke, which rose to the ceiling and escaped to the outside world through a louvre in the open-timber roof. With no glass in the windows and a reliance on wooden shutters for protection from the elements, there was a fine balance between sitting in a warm fug of eye-stinging smoke or allowing piercing draughts through the upper part of the room. In wooden buildings with thatched roofs there was also the ever-present risk of fire.

With improving technology in the building of houses in stone or brick, it became possible to construct a chimney as an integral part of the wall, creating an enclosed flue to conduct the smoke upwards and out through the chimney stack. With the hearth now moved from the centre of the room to one side, it was possible to contain it within a fire-proof alcove backed with stone or brick, which substantially reduced the danger of fire. By the Elizabethan era an elaborately carved chimney piece had become a major ornamental feature in a room and an important status symbol. At Avebury Manor, the very large fireplaces in both the Tudor Parlour and the bedchamber above it would undoubtedly have been a source of great pride to their owners.

ABOVE: In the Tudor period, flues and chimney stacks became major ornamental features. At The Vyne, in Hampshire, the brick flue is decorated with lozenge-shaped 'diaper work', picked out in blackened bricks.

LIGHTING IN TUDOR TIMES

The Tudors employed various methods for lighting their homes. Rushlights were an archaic form of domestic illumination, easy to make and economical. The rushes were picked in summer, soaked, peeled and bleached, then dipped in molten animal fat and dried. A single rushlight was held firmly at an angle in a metal holder (shown to the left of the picture) and gave off a steady light, but vigilance was required as the rush had to be moved along regularly and it dropped hot fat. Wealthier households preferred hand-dipped candles, made of tallow (pork or mutton fat, which smelt of the farmyard) or more fragrant but expensive beeswax. Families of distinction, such as the Dunches, might use candles in formal rooms, but behind the scenes the lowlier tasks were still conducted by firelight and rushlights.

Interiors and decoration

The monochromatic interiors we see today in Tudor homes are deceptive. The sixteenth century saw a general increase in wealth, and Henry VIII, in contrast to his frugal and cautious father, delighted in magnificent display. This mood of opulence influenced the new nobility, who decorated their houses accordingly. They revelled in colour and delighted in pattern – many surfaces were decorated, even if only with a simple pattern of studded nailheads. Oak panelling might well be painted in single, rich tones, perhaps decorated with coats of arms, and decorative plasterwork might have their details picked out in riotous shades. The predominant colours were those associated with heraldry: rich and strong colours such as crimson, azure, deep greens and acid yellows, with silver or gold highlights.

In Elizabethan times a new room type, the long gallery, became popular. Designed for indoor exercise and a display of wealth, long galleries were resplendent with paintings, coats of arms, family emblems, tapestries or embroidered wall hangings. In the past the wealthy had left their most sumptuous clothing to the Church, where the garments were unpicked and remade into copes and altar frontals, but the split with Rome put an end to this practice and so luxurious fabrics and embroideries were reused by the family to make cushions and hangings for the domestic interior.

TUDOR EMBROIDERIES

The art of decorative needlework is nearly as old as the skill of weaving, and luxurious English embroidery, known as the *Opus Anglicanum*, was famous across medieval Europe. After the Black Death wiped out one-third of the British population, however, including skilled artisans and the higher clergy who were their clients, demand for intricate embroidered robes declined. Embroidery was confined to being a holy endeavour for nuns and a leisure pursuit for women until the Tudors revived it as a profession, with court embroiderers such as William Ibgrave producing breathtaking garments for the royal family, even stitching jewels into the design. Embroidered items became appropriate gifts among the very highest, and in 1532 Henry VIII gave his great rival Francis I of France a red velvet bed embroidered with pearls. In 1562, the all-male Broderers Company gained its Royal Charter from Queen Elizabeth, a woman who appreciated the psychological impact of sartorial magnificence in an uncertain age.

ABOVE: This octagonal motif forms part of the embroidered panels made by Mary, Queen of Scots and Bess of Hardwick in 1569-84, during Mary's imprisonment. The designs were taken from woodcuts in contemporary books on natural history.

Embroidery in the home

Domestic embroideries often featured naturalistic patterns of plants and flowers copied from herbals and illustrated books. The most ambitious embroideries were those for bed hangings and matching coverlets. The needlewomen would first produce separate panels of canvas or linen, or appliqué small, sometimes tiny, embroideries worked on open-weave linen onto lengths of fabric, which would then be sewn together. Popular patterns featured the Tudor Rose, coats of arms and symbols of fertility.

Linen was favoured for domestic embroidery, and blackwork was extremely popular for personal apparel, with fine patterns worked in black silks on white linen; blackwork can often be seen in Tudor portraits. For interiors, there were many fabrics available, from simple

woven wools to sophisticated patterned velvets and damasks. Metal threads were used with embroidery silks, sometimes with small metal sequins (often called spangles). Luxurious dress fabrics were often reused in interiors to make cushions and hangings; after the ban on the wearing of ceremonial robes by clerics, resourceful needlewomen were quick to buy up and recycle the most magnificent copes and vestments for secular purposes.

Yet, while embroidery had grown into a craft practised at all levels in the Elizabethan era, by the end of the sixteenth century it had become either the preserve of the very wealthy, who could afford to commission the best, or the leisure pursuit of the less well-off who used their needle-wielding skills to decorate their domestic textiles and clothing. To cater for the latter, by the 1620s publishers were producing pattern books specifically designed for embroidery and drapers sold lengths of cloth complete with transferred patterns.

ABOVE: Embroidered by Bess of Hardwick and bearing her initials, this velvet panel was sewn using gold and silver thread.

BELOW: The intricately worked pear tree on this cushion cover is thought to symbolise Bess of Hardwick's children, the broken branches representing those who died in infancy.

HERALDRY

Heraldry evolved from the need to identify allies on the battlefield. Knights adopted distinctive motifs worn on fabric tabards over their armour (known as 'a coat of arms'), and their footsoldiers copied their 'team colours'. The coat of arms became a hereditary device in the twelfth century, awarded by the monarch, and began to appear as decoration for walls, tapestries and stained glass, further emphasising a family's elevated stature. Each unique coat of arms records a particular family's history, often combining mythical beasts such as dragons, gryphons and unicorns with more prosaic representations of boars and crows. The coat of arms (right) is that of Sir Francis Drake from Buckland Abbey, Devon.

ABOVE: Flemish tapestries, dating from around 1700, hang in the withdrawing chamber at Hardwick Hall.

Textiles and wallcoverings

Most tapestries and pile carpets were imported; although England had its own textile industry the majority of luxurious fabrics such as velvet, silk and linen damask were brought in from the Continent by merchants and drapers based in London. The indigenous textiles such as worsted and woven wool and everyday linen and hemp, with which most domestic furnishings were made, were sold routinely in larger towns.

Wall hangings of all kinds helped to keep interiors warm. The less expensive fabrics, such as canvas or buckram, were hung in the more modest home; these materials took hand-painting very well, imitating tapestry. Sequences of hanging panels of woven woollen cloth in strongly dyed colours, with a decorative border, were also popular. Such hangings would be suspended from the wall, absorbing sound and adding an air of luxury. For the very wealthy, there were embroidered wall hangings that fulfilled the same purpose but also demonstrated the prestige and affluence of the occupants of the house.

Oak panelling

Oak panelling was popular in Tudor households. Oak is a very hard, heavy wood and it was the predominant building material for secular structures until the sixteenth century. According to Bill Bryson in *At Home: A Short History of Private Life*, a typical fifteenth-century farmhouse consumed the wood of 330 oaks. In the sixteenth century, better-off households employed carpenters to fit out their rooms with rectilinear panels of carved oak, known as wainscotting, for extra insulation. A scheme of simple oak panelling, running from the floor to a plaster frieze below the ceiling, might be 'topped' with a decoratively carved, horizontal wooden border, as in the Tudor Bedroom at Avebury Manor. For those in search of more sophistication, architectural motifs such as flat columns or pediments could frame doorways, while linenfold panelling was always a popular choice.

One advantage of oak panelling is that it can be dismantled easily and is, therefore, surprisingly portable. The panelling in the two Tudor rooms at Avebury, for example, was purchased locally and installed in the early twentieth century by the Jenners, early proponents of architectural salvage. The complete oak panelling from the parlour at Sutton House – a remarkable survivor from Tudor times sited in Hackney, London – was illicitly removed overnight and put on sale. Fortunately the theft was discovered and the panelling retrieved and reinstated.

ABOVE: Linenfold panelling lines the dining room walls at Godolphin, a Cornish house dating from the Tudor and Jacobean eras. Of fifteenth-century Flemish origin, the linenfold motif was popular with Tudor wood-carvers for both panelling and furniture.

New Money at Avebury Manor

William Dunch (*c.*1508–1597) successfully navigated the political maelstrom of the Tudor era – serving two kings and two queens – and emerged from obscurity through the acquisition of wealth, land, influence and a coat of arms.

Under Henry VIII, William was a minor civil servant working for the Royal Household, recording the costs of events as diverse as Jane Seymour's funeral and Prince Edward's christening. In 1546 he became 'Auditor of the exchange of the coinage and mint of gold and silver coins and bullion' at the Royal Mint, receiving a salary of £50 per annum, which later increased to £133 a year.

BELOW: The family tomb in Little Wittenham church, Oxfordshire, shows William Dunch and his wife Mary at prayer. William died in 1597 at the impressive age of 89.

He rose through the ranks during the reign of Mary and received a Grant of Arms during the reign of Edward VI, then served as a Justice of the Peace for nearly 20 years and as MP for Wallingford. By 1570 he was High Sheriff of Berkshire, Queen Elizabeth's official representative for the county.

William was a successful and busy man, travelling the country on horseback or by coach, attending to his various business interests. However, the mid-sixteenth century, with royal power struggles and those between Protestantism and Catholicism, was an extraordinarily turbulent time for men of affairs. Incarceration was an occupational hazard and William was held in prison for short periods during 1549; he may have been implicated in the debasing of the gold and silver coinage, a policy which had been implemented by a cash-strapped Henry VIII shortly before his death.

There was another possible factor in a later imprisonment; the Dunches maintained a connection with 'Protector Somerset', the title of Edward Seymour, who in 1547 ruled in the name of Edward VI during his minority. Dunch was bailed from the Fleet prison for the vast sum of £4,000, but apparently pardoned in 1552, after Seymour was removed from office and executed on a charge of treason.

ABOVE: A fifteenth-century manuscript depicts the Tower of London, where a mint had been established some two hundred years earlier. It was here that William Dunch would have worked as 'Auditor of the myntes' from 1546.

Building a property portfolio

In 1546 courtier William Sharington had been put in charge of the Bristol Mint by Henry VIII, and he quickly accrued sufficient wealth to buy the Avebury estate in 1548. Arrested and charged with fraud the following year, he had to sell the property to meet his costs. William Dunch bought the estate in 1551 for £2,000, giving up his own role as Auditor of the Royal Mint to concentrate on acquiring land and having a family; in 1547, at the age of 39, he had married Mary Barnes with whom he had two sons. By 1557 William had built Avebury Manor on his recently acquired lands.

ABOVE: Charney Manor in Oxfordshire is one of the many houses once owned by William Dunch. This Grade I listed building dates back to the late thirteenth century and is now owned by the Quakers.

By the time of William's death in 1597, aged 89, the family were major landowners in Wiltshire and beyond. They rented out some of their farms and produced grain, livestock and timber on their own estates, employing local labourers. Perhaps surprisingly, William seems to have deliberately acquired land other than Avebury that included interesting and historic sites; Little Wittenham, in Oxfordshire, contains an Iron Age hill fort (thought in Tudor times to be a Roman site), while Ickford in Buckinghamshire has earthworks.

William and Mary Dunch were worldly and well-connected, with a thriving, comfortable household. Most of their staff were men, even in the kitchen. Female servants usually worked only as personal maids and nurses at this time since householders preferred to employ men to help deter robbers – this was still a dangerous age. Social divisions between the family and their servants were more nebulous in this era; indeed, well-off Tudor families often took in and trained the children of their poorer relatives, providing them with an education in return for gentle duties, 'waiting' on the family.

William seems to have deliberately acquired land that included interesting and historic sites.

A servant who worked for the Dunches and acted as under sheriff to William's son Edmund (who was appointed High Sheriff of Berkshire in 1587) was consequently able to describe himself as a 'gent'.

William's will

William Dunch always maintained good relations with the powerful and, in his will, his first legacy was to Queen Elizabeth:

'Under Almighty God I stand most bounden to the Queen's Most excellent Majesty first for that I was a sworne servant to her most noble father & to her brother & sister as also to herself & hath received good benefit and great princelie fauos [favours] from them &c. my executors to provide for her highness a ring with a diamond or peece of plate of value of £40.'

Ever the courtier, William also left silver cups worth £10 each to Sir Thomas Egerton, Lord Keeper of the Great Seal of England, Lord Burleigh, Lord Treasurer, and Sir Robert Cecil, Principal Secretary to the Queen. His land was split between his widow, his surviving son, Edmund, and other family members, while his impressive collection of silver was divided between his widow and his son to be used as capital for the rest of her life. Silver in the Tudor era was a liquid asset; robust yet portable, there was little sentimentality in selling it when circumstances dictated.

ABOVE: Part of William's estate at Little Wittenham included the twin hills known as the Wittenham clumps, or Mother Dunch's Buttocks. One of the hills was an Iron Age hill fort.

BELOW: William Dunch's primary residence was Little Wittenham, where he paid for the village church to be extended.

THE TUDOR PARLOUR

The term 'parlour' comes from the French word *parler*, to talk, and evolved from the name of the room where monks were allowed to break their silence and converse. Parlours were a recent innovation in Tudor homes, where the benefits of individual rooms for specific functions had become apparent. The communal hall often survived as an entrance hall or dining room, but for relaxation the family would withdraw to the parlour where they could sit in greater comfort. It was here that guests could be entertained, business transacted, marriages brokered and deals done.

The Parlour at Avebury Manor

The parlour at Avebury lies in the south extension of the original Tudor house and was possibly built by James Mervyn in about 1600, along with the addition of the new south range. It is a substantial, well-proportioned chamber with stone-mullioned and transomed windows – large by the standards of the time – on three sides. The elegant plasterwork ceiling is executed in low relief, with an interlocking geometric pattern typical of the Elizabethan era, painted white to maximise reflected daylight.

BELOW: The Tudor Parlour during the Jenners' time was decorated in early eighteenth-century style. The wooden panelling was salvaged from elsewhere and installed by the Jenners.

The room has been used as a parlour throughout its history, though its appearance has been constantly modified to suit the changing styles of different periods; a photograph from the late nineteenth century shows the room typically cluttered with Victorian furniture and artefacts. The same photograph reveals that there was no wooden panelling at the time, but in the early twentieth century the Jenners installed old oak wainscoting, restoring the room to its original Tudor form. They also found the impressive stone fireplace in an outhouse and had it restored and reinstated.

Furnishing a Tudor parlour

While Tudor furniture often looks angular and uncomfortable at first sight, much of the wooden furniture surviving today would once have had cushions and textile coverings. In addition, the Tudor house would have been cold for at least six months of the year, so both men and women wore many layers of clothing which would have made sitting on a hard wooden bench more tolerable. Extra warmth was provided by textile panels on the walls; well-off families favoured expensive tapestries, while the more frugal used hand-painted textile panels.

ABOVE: The same room in the late nineteenth century during the Kemms' tenancy at Avebury. Note the contemporary fireplace and typical Victorian clutter.

BELOW: The elegant low relief plasterwork on the parlour ceiling is typical of the Elizabethan era and would have been painted white to reflect the light.

ABOVE: Cushions, such as this elaborately embroidered one from Hardwick Hall, Derbyshire, added comfort to the Tudor home.

Padded seat furniture, whether upholstered or in the more portable form of cushions, added greatly to comfort in wealthier households. Often the covers were made from ecclesiastical robes and copes, bought after the Reformation, or from luxurious clothing fabrics; some cushions were made of robust tapestry and backed with leather, to be used as seating on the floor.

At mealtimes, people usually ate at simple refectory tables, mostly sitting on robust four-legged 'joyned stools'. Important visitors and senior family members, however, tended to have chairs with backs, while the head of the household would have a chair with arms, known as an 'elbow chair'. This idea of social status still lingers in modern suites of furniture, where an even number of dining chairs are 'topped and tailed' by a pair of carvers at each end, to be occupied by the host and hostess. Other typical forms of furniture included the court cupboard:

THE AUMBRY

The aumbry, or dole cupboard, was originally a cabinet in the wall of a church or monastery where artefacts such as chalices were stored. In time, it became the place where left-overs from the tables of the wealthy were placed for the benefit of the poor – hence the phrase 'on the dole'. The charitable function of aumbries lapsed with the Dissolution of the Monasteries and, instead, the word came to mean the food cupboard of better-off families. The sides and top were sometimes pierced to allow air to circulate, as in this one at Selly Manor in Birmingham, in which case the interior would be lined with a robust fabric to deter vermin.

a type of low dresser, usually with an open shelf below and recessed cupboards above, used to display vessels, pewter and silver for dining. The aumbry, or food press, was another type of cupboard, with lockable doors in which food was kept away from vermin.

Opulent carpets were highly prized in the Tudor era. They were usually displayed on flat surfaces, such as tables, as they were far too precious to put on the floor, which until recently had been the repository for all manner of discarded matter. As late as 1530, the visiting philosopher Erasmus wrote to Cardinal Wolsey's physician, describing with disgust his visit to an English house where the rush-strewn clay floors were deep in domestic refuse. By the Elizabethan age, however, oak boards were often covered with a layer of woven rush-matting, sweet-smelling and warm underfoot.

BELOW: At Sutton House, London, the carpet is displayed on the table, as was common in Tudor times when carpets were considered too precious to walk on.

TUDOR TAPESTRIES

In the fourteenth and fifteenth centuries, tapestries were increasingly to be found gracing the interiors of grand European homes. As the manufacturing technique was lengthy, highly skilled and expensive, these large-scale hand-woven textile hangings became sought-after status symbols and towns such as Flanders in the Low Countries specialised in manufacturing them for export to the wealthy elite.

The designs, known as 'cartoons' (from the Italian word *cartone*, meaning sheet of paper), were created by artists drawing on heavy-duty paper. This would be cut into vertical strips which were placed behind the warp of the loom and used as a pattern by each highly skilled weaver. Working side by side the weavers

ABOVE: This detail of the tapestry in the Queen Anne's Room at Cotehele, Cornwall, depicts a bucolic scene with birds and deer.

BELOW: Tapestries depicting the story of Ulysses line the walls in the High Great Chamber at Hardwick Hall, Derbyshire.

followed and interpreted the pattern, linking their work with their neighbours' in a seamless progression as they built up the image. The materials used were vibrantly coloured silks, wools and metallic threads, and as the piece was woven from behind the design would be reversed in the finished tapestry.

As well as religious subjects, historical narratives and bucolic hunting scenes were much in demand; with the spread of Renaissance learning, classical myths and legends from Greece and Rome also became desirable. Tapestries often incorporated architectural features such as columns and arches to give a format to the scene, and these reflected the influence of illustrated books and prints of triumphal arches and classical ruins.

By the age of the Tudors, most of the aristocracy owned sets of tapestry hangings, which acted as sound bafflers and welcome insulation in chilly room settings. There was a lively trade in second-hand sets and it is not uncommon to find a 'cut down' tapestry, thriftily reused; they were often sold at a certain price per ell, a measurement of 68.5cm sq (27in sq). Those who could not afford woven tapestries with figurative scenes might buy a 'tapisserie de Bergame' – cheaper woven textile hangings featuring large-scale, repeating patterns. For the hard-pressed there was also a market in cloth hangings, where scenes were painted on linen scrim to imitate the effect of tapestry.

An indigenous industry

By the seventeenth century there was a thriving British tapestry industry, supported by James I and based at Mortlake, where Flemish workers trained English apprentices to produce superb textiles. Tapestries remained fashionable among the wealthy until the end of the eighteenth century, when they were largely superseded by the development of printed wallpapers and mass-manufactured woven textiles for wall-covering. However, the technique was revived with great success at Merton Abbey in 1881 by William Morris, and found new markets among the Victorian wealthy.

ABOVE: This border from a tapestry in the Tapestry Bedroom at Speke Hall, Liverpool, was woven at Mortlake *c.*1700.

Transforming the Tudor Parlour

While the Tudor Parlour at Avebury Manor is not the oldest room in the house it is the one that has received the earliest design treatment, placing it in about 1560. One of the key objects of this project was to reflect the characters who lived in this house – their life and times, their social status and their preoccupations. Through a combination of research and consultations with experts Dan Cruickshank and Anna Whitelock, it became apparent to designer Russell Sage that by the start of Elizabeth I's reign in 1558, the Dunches were an established *nouveau riche* family with good connections at Court, considerable wealth and property, and a brand new coat of arms awarded by the previous monarch, Edward VI. To reflect this, Russell and the experts created a design scheme that epitomises William and Mary Dunch's view of their place in life and their entitlements. By the standards of the day there is a notable air of conspicuous consumption, with newly commissioned carved furniture, impressive fixtures and class-conscious fittings. Anna humorously refers to 'Tudor Bling' to describe such ostentatious display as the painted heraldic panel and freshly-woven tapestries which would have indicated the family's prestige and money.

LEFT: The heraldic panelling in the Winter Parlour at Canons Ashby, Northamptonshire, was the model for the heraldic panel in the Tudor Parlour.

The Heraldic Panel

To reflect William Dunch's status, the team had wanted to follow the authentic Tudor practice of painting heraldic motifs directly onto the wooden wall panelling, but his proposal was vetoed by the National Trust's curatorial experts because it could damage the historic woodwork, and all aspects of this project had to be reversible. Undaunted, they came up with an ingenious alternative; Russell scoured the country to find an appropriate section of panelling that could be decorated with the Dunches' coat of arms and then hung *in situ* above the fireplace.

Russell commissioned fine decorator Grant Watt to paint the heraldic panel. His design was based on the heraldic painting in Canons Ashby's Winter Parlour, a colourful scheme dating from the 1580s that celebrates the Dryden family's ancestry, with various family emblems painted directly onto the walnut panelling. The design for Avebury, created with the specialist assistance of the College of Arms, incorporates both William Dunch's original coat of arms, which he acquired in 1550, and a subsequent version based on the coat of arms of one of his descendants. It is interesting to note the difference between the two designs; dynastic marriages, service to the Crown and awards often changed the appearance of a family's coat of arms beyond recognition in just a few generations, a further incentive to display all coats of arms to indicate an honourable lineage.

BELOW: Fine decorator Grant Watt gilds the heraldic panel at Avebury.

Crafting the Furniture

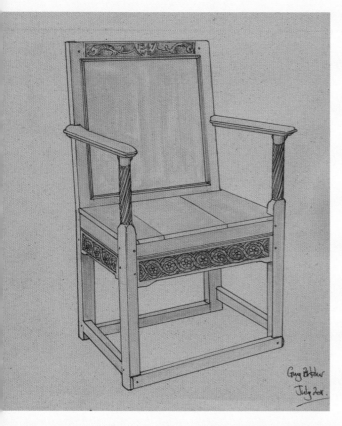

To create the type of interior that a high-status family like the Dunches would have used, Russell and the experts drew on historic examples from the era. Extensive research informed every design decision, and the best possible makers and craftspeople brought their experience and knowledge to the project.

To craft the furniture, Russell turned to Guy Butcher, a furniture designer and maker with a workshop near Ledbury, Herefordshire. He uses predominantly native hardwoods and is fascinated by historical styles and construction techniques. He made all the oak furniture for the Tudor Parlour – a joined table, two joined chairs (his and hers), four joined stools (two single seats and two doubles) and an aumbry – using modern tools but traditional techniques that would have been employed around 1560.

ABOVE: A design visual for William Dunch's chair depicts the carved cresting rail and guilloche pattern around the seat, which were finely carved by master carver Emyr Hughes.

BELOW: This detail of the cresting rail of William's chair shows the fine carving, including William and Mary's initials and the date of their marriage.

The table is a simple refectory style with chamfered legs and a cleated top, typical of the period. 'His' chair has a planked seat, carved seat rails and a cresting rail with Renaissance-influenced carving, containing the initials of William and Mary Dunch and the date of their wedding, 1547; 'her' chair is lighter in construction and is a design known as a 'caqueteuse' or 'gossip's' chair because the seat narrows to the rear, making the user sit forward slightly, as though sharing a secret. The joined stools were made as a matching set and have turned and fluted legs, with some relief carving along the seat rails. The aumbry is a simple planked construction with two doors and is pierced with gothic tracery.

All the pieces were drawn from authentic historical examples. The aumbry is based on one in the Cadbury Collection at Selly Manor Museum, owned by the Bournville Trust; the caqueteuse chair is in the V&A, while Sir William's chair was based on a well-documented piece made before 1527 for Sir Rhys ap Thomas of Carmarthenshire, a man closely connected to the Tudor court. Guy's models for both chairs reveal a slight French or Flemish influence, since this was an era when itinerant Flemish furniture-makers settled in England and Wales.

ABOVE: Mary Dunch's chair is known as a 'caqueteuse' or 'gossip's' chair.

ABOVE LEFT: Cabinet-maker Guy Butcher finishes carving the spiral twist for the arm of William's chair.

The wood used for all the furniture is seasoned oak, finished with a little beeswax to give a protective finish and a natural colour. Most of the pieces are of joined construction, using mortice and tenon joints, many of them reinforced with oak pegs, made exactly as they would have been in Tudor times. The few metal components – the hinges, nails and lock on the aumbry – were made by a local blacksmith, copying authentic historic forms. Some wooden components were sent to be turned by hand on a traditional pole lathe; others were carved by master carver Emyr Hughes, who was particularly knowledgeable about the techniques and styles of the period.

ABOVE: The mortice and tenon joints of the stool are pulled together and held in place by oak pegs, which is a process called braw boring.

BELOW: Presenter Paul Martin gets to try his hand at pole lathing under the watchful eye of green woodworker Gudrun Leitz, who collaborated with Guy on making the Tudor furniture.

MAKING THE RUSH MATTING

Rush matting is an ancient material that has been used to cover floors since the Middle Ages. It provides a finished flooring that is both resilient and appealing, warm underfoot and attractive for its variable colours and intricate texture. It also gives off a pleasant aroma that is refreshed by sprinkling the surface with water.

Specialist makers Rush Matters, headed by Felicity Irons, were commissioned to provide authentic rush matting for the Tudor Parlour. The process begins with the harvesting of English freshwater bulrushes (*Scirpus lacustris*) from the Great Ouse River, in Bedfordshire, between June and August. Using 5m- (17ft-) long punts to reach the most inaccessible clumps, the rush cutters cut the bulrushes from the river bed with slender scythe-shaped knives attached to long handles to obtain lengths of up to 3m (10ft). The stems are then dried in the sun and wind, after which they are plaited by hand into continuous lengths 7.5cm (3in) wide. Lavender and southernwood, popular in Elizabethan times for their moth-repellent qualities, are also threaded into the plaits. The strips are then sewn together with jute twine and the edges bound with a separate plait, 4cm (1½in) in width, to form individual mats, each one purpose made to fit a particular room.

ABOVE: The dried rush stems are plaited together.

BELOW: Designer Russell Sage lends a hand gathering in the rushes one damp summer morning.

Tudor craftsmen

Tudor craftspeople learnt by instruction and imitation, assimilating designs and construction techniques as they came across them. They might keep their own sketch books and have recourse to printed pattern books, and would have encountered foreign craftsmen through trade and travel. Certainly Continental Renaissance influences were becoming evident in English furniture by the 1560s, as people and ideas moved around during this relatively peaceful and prosperous period. However, while people were mobile, timber was expensive to transport any great distance, so local materials would have been sourced. Guy believes that the furniture at Tudor-era Avebury Manor was probably made no farther afield than Salisbury, at that time a regional centre for furniture-making, and the less sophisticated pieces might even have been made at Avebury itself, by itinerant craftsmen using seasoned timber from the Dunches' own estate.

Making the Tapestries

It would have been prohibitively expensive to make new tapestries, but a technologically advanced company offered an alternative solution based on more affordable hand-painted Tudor wall hangings. Zardi and Zardi specialises in reproducing historic textiles, from embroideries and woven fabrics to tapestries. The firm was commissioned to source and print

BELOW: The Abraham Tapestries at Hampton Court Palace provided the inspiration for Zardi and Zardi's wall hangings in the Tudor Parlour.

onto fabric images of tapestries taken from appropriate historical references. Russell and Dan discussed the possible sources, and the firm settled on a scheme of six pieces that are replicas of the Abraham Tapestries from the Great Hall at Hampton Court. The originals were commissioned by Henry VIII and portray scenes from the Old Testament.

Photographs of the Hampton Court tapestries were digitally manipulated so that the proportions could be adjusted to fit into the Tudor Parlour, and each design was printed on to a piece of heavyweight linen, woven in Scotland. Although smaller than the originals, they are still impressive in size; each panel is 3m (10ft) high, and the width varies from 1.2m (3ft 10in) to 4.5m (14ft 10in). There was a debate about the strength of colour to be chosen for the new pieces: should they look as they would have when just made, more than 450 years ago, or should they be more mellowed? Bearing in mind the exposure of tapestries to variable daylight from three angles, which would have caused fading, combined with atmospheric pollution from the fire's wood smoke and primitive forms of lighting, it was felt appropriate to use the more muted colour palette of the Hampton Court tapestries.

ABOVE: Simon Nagy of Cadworks overseeing the printing of a tapestry panel for Zardi and Zardi.

The Finishing Touches

Throughout this project Russell and the experts were keen to evoke the authentic texture of everyday life by including appropriate details and props wherever possible. He finished off the Tudor Parlour with pewter vessels found in sale rooms and at auctions, modern ceramics made to historical designs, and sycamore trenchers turned by Guy Butcher. There are no window curtains in this room, as they would have been a rarity in country houses of this era, but authentic rushlights and candles give an accurate impression of manor house life in the sixteenth century.

OVERLEAF: The finished room is designed to reflect the high social status of William and Mary Dunch, complete with heraldic panelling, tapestry-lined walls and newly-made furniture.

FROM AVEBURY TO THE NEW WORLD

I n 1582 Walter Dunch took over Avebury Manor and its estate from his father, William. However, Walter predeceased his father in 1594, leaving his widow Debora to care for his four daughters and one-year-old son and heir, named William after his grandfather.

The following year a dispute arose between Debora and a neighbour, Richard Truslowe of Trusloe Manor, over the ownership of a sixteenth-century pigeon house, which still stands close to Avebury Manor. By law, only the lord of the manor was entitled to keep pigeons, which were a useful source of fresh meat. By laying claim to the pigeon house, Richard Truslowe was therefore attempting to usurp the prestigious title of lord of the manor.

ABOVE: The sixteenth-century pigeon house was at the centre of a dispute between Walter Dunch's widow, Debora, and a neighbour, Richard Truslowe.

BELOW: Debora and Sir James Mervyn's elaborate tomb in Little Wittenham Church in Oxfordshire reflects the high status of the family by the early years of the seventeenth century.

The legal tussle was neatly resolved when Debora Dunch married the influential Sir James Mervyn of Fonthill Gifford, an MP and the High Sheriff of Wiltshire, who swung the verdict in his wife's favour.

Sir James continued the expansion of Avebury Manor, extending the original house southwards and building a new south range, with its Renaissance-style doorway and plaque bearing the initials MJD, thought to stand for Mervyn, James and Debora.

An indomitable daughter

Walter and Debora Dunch's daughter Deborah was a remarkable individual who founded an early colonial settlement in North America. She married Sir Henry Moody and they lived near Malmesbury in Wiltshire until Sir Henry died in 1629. A committed Anabaptist, believing that children should not be baptised until they could appreciate the meaning of it, she was persecuted for her beliefs so in 1639, at the age of 53, she moved to Massachusetts, taking her son Henry with her. English Puritans had settled along the eastern seaboard of America so that they could freely follow their religious convictions. However, Lady Moody's beliefs were too trenchant even for the Puritans, so she joined the Dutch settlers, and was promptly excommunicated.

Under her leadership, the group founded a settlement, Gravesend, in New Netherland in 1645 (now part of Brooklyn in New York City), the only permanent settlement in America's early colonisation period to have been designed, planned and directed by a woman. Lady Deborah's travails included attacks from indigenous Americans along the Hudson River and vilification by John Endecott, the English Deputy-Governor, who described her as 'a dangerous woeman'. Nevertheless, she was determined to negotiate a workable and peaceable relationship between the Dutch and the English settlers in the New World, free from religious oppression, despite the war which broke out between Holland and England in 1652. She died in 1659 aged 73 years.

ABOVE: A plaque bearing James and Debora Mervyn's initials was placed above the entrance porch of their newly completed south range.

The Tudor Bedchamber

This is a particularly attractive room situated on the first floor above the Tudor Parlour, dating from around 1600, with an ornate plasterwork ceiling and a deep decorative moulded plaster frieze above the wainscoting. Each of the motifs in the ceiling would have had considerable significance, the peacock, for example, symbolising immortality and heaven. The oak panelling, although largely seventeenth century in origin, was installed by the Jenners when they restored the manor.

The handsome carved stone fireplace is original, while the overmantel panel is made of plaster or fibreglass and is also a twentieth

ABOVE: This photograph shows the Tudor Bedchamber as it was in the early twentieth-century, furnished in period style. The ornate plasterwork ceiling dates from around 1600.

BELOW: Expert Anna Whitelock inspects the recently restored fireplace, which is original to the room and is carved with some fine classical details.

century addition. A small, discreet anteroom leads off the Bedchamber. Rooms of this sort provided householders with a place for washing and bodily functions. Judging by the pattern of the ceiling, the anteroom was created before the plasterwork was applied.

Tudor bedchambers

The bedchamber evolved in the mid-sixteenth century as a specific room for slumber, although they also served as public rooms. Previously people frequently slept wherever space allowed, and in some Tudor houses the parlour boasted a bed as a permanent fixture, an early form of bedsitter; servants meanwhile regularly bunked down on portable straw mattresses. In his *Book of Household Rules* (1595), Lord Montague wrote of his servants: 'every morning they doe ryse att a convenient hower to remove the palettes (if there be any) out of my said withdrawing chamber'. Alternatively, servants slept on a 'truckle' or 'trundle' bed, a simple wooden frame with no head or foot board, mounted on short legs with castors or wheels. It had a lattice of stout woven ropes or webbing as a support for a straw mattress, topped with a blanket. The truckle bed slid inconspicuously under its larger neighbour during daylight hours.

'Every morning they doe ryse att a convenient hower to romove the palettes out of my said withdrawing chamber'.

PRIVATE ABLUTIONS

For all classes of people, rudimentary indoor lavatory facilities were provided by the chamberpot, a solidly made, single-handled receptacle known as a 'jordan'. Portable, robust and versatile, each jordan was emptied several times a day in the 'house of office', a modest wooden structure enclosing a deep pit, over which was a bench with one or more holes. Generally built adjacent to the back of the house, the house of office was often designed to accommodate several users simultaneously, in cheery proximity. More affluent householders might own a close stool, a portable piece of furniture rather like a solid chair with a hole in the seat, under which was placed a chamberpot containing water, to be removed and emptied by servants after use. The one featured here is from Hampton Court Palace.

ABOVE: Lavish bed hangings and tapestries, such as the ones seen here in the Green Velvet Room at Hardwick Hall, Derbyshire, were extremely expensive and as such were symbols of great wealth in the Tudor era.

ABOVE: In the past it was common practice to reuse salvaged parts from older pieces of furniture. This bed post at Cotehele, Cornwall, once formed part of an Elizabethan table.

Beds could be valuable assets; in well-off households, 'best beds' were left empty to accommodate esteemed visitors, while the householder and spouse occupied 'the second-best bed'. As Anna Whitelock observed, 'Beds and sleeping arrangements provide a fascinating means of understanding domestic politics throughout history; where and how individuals slept tells you a great deal about society, and people's relative positions within it.'

Typical beds were four-poster frames of oak, heavily carved, with an ornate headboard and two end posts, linked by a tester, or canopy. The headboards often sported a niche for a candle, and the decorative carvings were biblical or classical in inspiration. The frame of the bed had holes bored at regular intervals to house a network of ropes which needed frequent adjustment to keep the surface taut and to prevent the occupants rolling together (hence the archaic phrase 'night, night, sleep tight').

Bed hangings were appropriately lavish for the most expensive piece of furniture in the house. Curtains hung around the outside of the bed, while narrow panels filled the frame, providing warmth and privacy; a four-poster could be as snug as the enclosed box-bed of previous centuries. The hangings were often richly embroidered in wool-work, stump-work and appliqué, perhaps with bucolic scenes or abstract floral designs. The mattress would have been made of straw, the term having derived from 'matted truss', the practice of gathering and bundling straw in a large fabric sack on which to sleep, usually in some discomfort.

Other essentials

Householders traditionally kept their valuables, bed linen, clothes and personal belongings in lockable rectangular coffers or small chests, combining security with portability. By the end of the Elizabethan era a simple wooden chest had evolved, with a jointed frame held together internally with pegs. The sides of the chest were made from individual panels, carved and assembled into a whole, rather than from planking. Inside, there might be a shallow shelf for storing candles.

Typical beds were four-poster frames of oak, heavily carved, with an ornate headboard and two end posts, linked by a tester.

Box chairs, too, were popular with wealthy merchants and the nobility. The base was panelled rather than open, so these chairs offered protection from floor-level draughts and supplied some comfort in bedrooms, particularly when padded with cushions. The fore-runner of the modern armchair, they provided a seat with high sides and solid arms to help the sitter rise and descend with little effort. The back was often carved with decorative motifs or linenfold panelling.

LEFT: Carved wooden chests, like this one at Bradley Manor, Devon, provided storage in an era when clothing was often too heavy to hang up and too expensive to leave unprotected.

TUDOR PLASTERWORK

ABOVE: A beautiful example of a Tudor plasterwork ceiling can be seen in the Great Hall at Buckland Abbey, Devon.

The Tudor Bedroom at Avebury has an attractive plasterwork ceiling that dates from about 1600, when this part of the house was built. The ceiling is decorated with a proliferation of moulded plaster motifs which are not only decorative, but were chosen for their symbolic value. To Elizabethans, the peacock represented immortality and heaven; the lion embodied personal attributes such as courage and strength; and the unicorn alluded to the classical figure of Artemis, goddess of wisdom, as well as having Christian significance and associations with spiritual purity. As a more pragmatic mark of respect to the monarchy, the ceiling design also incorporates a Tudor rose.

Throughout the sixteenth century moulded lime plaster ceilings and friezes were created by teams of itinerant plasterers, many of them originating in the West Country. A plasterer could offer a large range of designs to a client; he usually travelled with a small portfolio or sketchbook of suitable designs, some his own, others copied from pattern books, to be adapted to fit rooms of most dimensions, or he might devise an exclusive design for a particular client.

One fascinating survival from a family of plasterers, the Abbotts of North Devon, is a small pocketbook bound in leather, now in the Devon Record Office. It contains more than 300 designs which were used by the family between about 1550 and 1700. The earliest patterns featured in this book, in careful ink drawings, are very similar to the parlour ceiling at Avebury: subtle, geometric repetitions of interlocking forms, with some curves.

The creation of a plaster ceiling was a lengthy business. A sticky lime plaster mixture was made up, sometimes reinforced with goat or horse hair, and poured into carved wooden moulds, which would be reused where possible. Once the plaster was dry, the moulds would be turned out and the low-relief forms fixed *in situ* to the ceiling lathes to form a

continuous low-relief pattern; gaps and joins would be touched up by hand. A plaster ceiling typically took a year to dry out completely, before being limewashed. Ceilings were traditionally painted white to to help reflect light back into the room, as every last ray of sunlight was precious.

Some contemporary ceilings were notable for their patriotism; Lytes Cary Manor has a magnificent coved and ribbed plaster ceiling in the Great Chamber, dating from 1533, the focal point of which incorporates the coat of arms of Henry VIII. Later ceilings tended to be deeper in relief and more figurative, with designs of people and animals and pendant forms of great complexity.

ABOVE: This symbolic unicorn is a feature of the plasterwork ceiling in the Tudor Bedchamber, and is associated with spiritual purity and wisdom.

BELOW: Tudor coved plasterwork ceiling in the Great Chamber at Lytes Cary Manor, Somerset, dates from 1533 and incorporates the coat of arms of Henry VIII.

TRANSFORMING THE TUDOR BEDCHAMBER

The inspiration for this room was the union of Debora Dunch, widow and owner of Avebury Manor, with the influential Sir James Mervyn, an MP and the High Sheriff of Wiltshire. The theme thus reflects the institution of marriage and the mingling of two well-placed and affluent households; Debora's involvement with powerful Sir James effectively resolved her legal dispute with a neighbour, while he acquired a desirable country manor house and a ready-made family.

With this premise in mind, there were a number of key design decisions to be made; should all the furniture be newly made, or original pieces of the correct vintage found? Would the plasterwork have been decorated, and if so, how bright would the colours have been? What sort of textiles would have been used for the bed hangings, and was it possible to re-create them using authentic methods? The solution reached reflected the fact that original Tudor furnishings would have been hard to find; instead, a period bed was renovated and traditionally woven textiles were commissioned for the bed hangings, combining old with historically accurate new.

BELOW: The Tudor Bedchamber is seen here before its transformation. The oak panelling is seventeenth-century, and was installed by the Jenners when they restored the property.

Restoring the Bed

Designer Russell Sage acquired an early seventeenth-century bed at an auction, of a date probably later than the creation of this room, though authentic in style, materials and construction. It is made of oak, with a headboard of twin arched sections richly carved with stylised plants and architectural details. The carving combines winged cherub heads and a horizontal band of dragons or serpents. This venerable survivor needed some attention; the condition report specified 'marks, scratches and abrasions consistent with age and use. Old splits . . . sections of replacement and losses'. The bed was renovated to consolidate all its components, and the structure reinforced to make it more robust.

While working on the bed, specialist furniture restorer Dave Lyons found that no glue had ever been used in its construction; it was entirely held together by dowel pegs fixed through the joints. A less welcome find was the furniture beetle, or woodworm, that had infested parts of the front legs, which meant planing back the crumbling wood and replacing it with new oak pieces.

After consulting the experts, designer Russell Sage arranged for the bed to be turned into a four-poster, which would be more appropriate to a country manor at the turn of the sixteenth century. This involved extending the headboard and adding new turned and carved end posts and a moulded and carved wooden cornice and frieze at the top. Some of the carving echoes the meaning of the Green Man depicted in the plaster frieze (see page 94). There was some debate about how tall to make the four-poster – as the bed is only 1.38m (4ft 6in) wide it was important that the height was in proportion and that the wooden cornice should fit below the plaster frieze in the room.

ABOVE: The bed is made up of a mixture of old and new parts. Here, Stephen Edwards of the Four Poster Bed Company, turns one of the urns for the bed posts.

The headboard was extended using early carved-oak caryatids from an old panel acquired by Russell, and additional moulded and turned elements were made by Stephen Edwards at the Four Poster Bed Company, a specialist maker of bespoke beds in historic styles. Once the details were supplied and agreed with expert Dan Cruickshank, some pieces were sent to a specialist carver, and once returned they were finished and glued where necessary. To support the mattress, Stephen created an authentic rope lattice, running from top to bottom then side to side, through existing holes in the frame, pulled taut to create a resilient open mesh. The bed was delivered to Avebury Manor in parts (though not on a horse and cart, which would have been taking authenticity too far) and constructed *in situ*.

Weaving the Bed Hangings

The bed hangings and textiles were vital to the re-creation of an authentic bedroom of about 1600, and Ian Dale, the brains behind Angus Weavers of Brechin in Scotland, was the natural choice to provide all the bed fabrics for this room. The firm is the last remaining maker of Jacquard linen in Britain and it reproduces historic textiles for clients as diverse as Dover Castle, Hampton Court, Kew Palace and the National Trust, as well as bespoke commissions. The handlooms used have been operational in Scotland since the 1760s and the skills employed are even older. Ian is a Master Weaver and a former Deacon of the Aberdeen Weaver Incorporation, which received royal approval in 1380.

BELOW: Ian Dale, of Angus Weavers, is seen here weaving the bed hangings on an eight-panel loom frame dating from the 1760s.

Ian carried out extensive research in the vast textiles archives at the Victoria and Albert Museum and devised a range of authentic fabrics and colour schemes to be hand-woven on treadle-looms for the bedchamber. The bed hangings and tester pelmet are made of say – a heavy linen twill – trimmed with decorative braid and a fringe. Hand-woven say is a very handsome fabric, made from flax imported from Flanders, which imparts a naturally brilliant, shimmering quality. While very wealthy Elizabethans might have preferred more showy, opulent fabrics, say would have been an excellent choice for a newly married country couple of means.

The coverlet is of hand-woven linen twill, a ground often used by Elizabethan ladies for crewelwork embroidery. Motifs known as 'slips' were worked separately by students and staff at the Royal School of Needlework. They were embroidered in crewel wool, each one depicting the 'eye' found in a peacock's tail feathers; the design and colouring were based on a detail in the room's ceiling. Authentically following the technique of the period, each slip was then appliquéd separately to the coverlet.

The bed hangings have linen linings, and the bedding itself has specially woven coverings for all its components. The three mattresses (of straw, wool combined with horsehair, and duck down), as well as the bolsters, pillows and sheets, are covered with either linen or ticking (so called because its very tight weave deterred the passage of ticks). There is also a woollen blanket, Wiltshire being famed for its wool at the time. Even the ropes supporting the mattress have been made from high-quality flax.

Painting the Plasterwork

There was much debate about decorating the Tudor Bedchamber's plaster frieze and ceiling. Although the plasterwork in this room has only ever been rendered in white, there are historical precedents for vividly painted moulded ceilings from this era. Referring to the decorative scheme at Plas Mawr in Conway, it was agreed that the moulded plaster frieze above the panelling – with its foliate forms and figure rising from what looks like a pot – should be

ABOVE: Expert Anna Whitelock embroiders one of the slips at the Royal School of Needlework.

painted. The figure was identified as the mythical Green Man of pagan times, the ancient spirit of nature representing fertility and plenty, so it was thought appropriate to use a range of shaded and high-lit greens, with the pot painted terracotta. After consultation with the College of Arms, details on the ceiling – the symbolic figures and the all-important Tudor Rose – were highlighted in stronger colours to emphasise the heraldic quality of the design. Fine decorator Grant Watt carried out the work using modern paints by Farrow & Ball and Earthborn Paints.

The Finishing Touches

Long, plain woollen hangings were used to block off the light from the windows; in country houses of this time, curtains were not needed in bedrooms as the bed hangings provided both privacy and warmth. As a finishing touch, a Tudor-style chest and two chairs were placed in the chamber, and the anteroom was provided with both a truckle bed – of the sort used by servants and specially made for this project – and a close stool, so the various needs of the bedchamber's occupants could be quickly met.

RIGHT: The Tudor Bedchamber is resplendent with its four-poster bed, combining old and new parts, and specially commissioned bed hangings.

BELOW: The details on the moulded plaster freize have been picked out in colour. The central figure is believed to represent the Green Man of pagan times.

CHAPTER III

AVEBURY IN THE 18TH CENTURY

THROUGHOUT THE EIGHTEENTH CENTURY THE APPEARANCE OF AVEBURY MANOR WAS INFLUENCED BY THE CLASSICAL STYLE, IMPORTED FROM ITALY AND BEYOND BY SUCCESSIVE OWNERS ANXIOUS TO IMPRESS THEIR SOCIAL SUPERIORS. THE NEWER SOUTH WING WAS TRANSFORMED BY THESE ENLIGHTENED IDEAS OF DECORATION, AND ARCHITECTURAL DETAILS THAT PROPERLY BELONGED ON RUINED TEMPLES IN THE MEDITERRANEAN WERE COPIED TO ADD A VENEER OF GRAND TOUR SOPHISTICATION TO A RURAL WILTSHIRE MANOR HOUSE.

LIFE IN THE AGE OF ENLIGHTENMENT

ABOVE: Classical sculptures line the staircase hall at Hatchlands Park – an eighteenth-century mansion in Surrey.

The eighteenth century was an era of increasing prosperity, with considerable trade between Britain and the New World. The East India Company, founded in 1600 to secure supplies of spices and exotic goods, now imported tea, coffee, cotton, silk, chintz and ceramics and was highly influential in expanding Britain's imperial ambitions, especially in India. On the international stage, relations between Britain and France were fraught, culminating in Britain's condemnation of the French Revolution and the rise of Napoleon. By the end of the century, the American colonies having gained their independence, the British government was now planning to exploit the vast, unknown continent of Australia.

Culturally, the Age of Enlightenment ushered in a respect for rationalism, scientific and philosophical progress and independent thought. The wealthy were smitten by the trappings of the antique world; keen to show off their sophistication, they advertised their passion for classicism by commissioning well-known architects to house them in Palladian mansions, studded with sculpture

18TH CENTURY TIMELINE

George I

1702
Queen Anne, last of the Stuart monarchs, accedes to the throne

1714
The Elector of Hanover – great-grandson of James I – becomes George I, Britain's first Hanoverian king

1727
George I's son accedes to the throne as George II

1706 AND 1707
Acts of Union unite the Scottish and English parliaments and 'Great Britain' is formed

***c.*1720**
Palladianism becomes the leading architectural style for country houses

ABOVE: Stourhead, in Wiltshire, is one of the first and finest Palladian mansions to
be built in England. It was designed by the leading Palladian architect Colen Campbell
in 1721 and was inspired by Palladio's Villa Emo (c.1564) in Fanzolo, Italy.

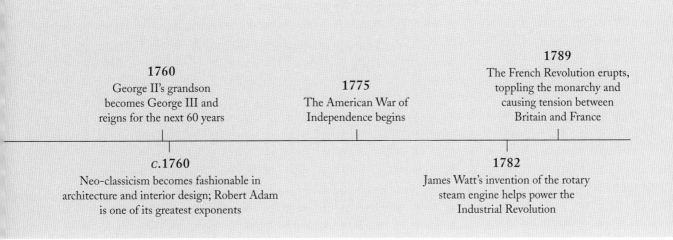

1760
George II's grandson
becomes George III and
reigns for the next 60 years

1775
The American War of
Independence begins

1789
The French Revolution erupts,
toppling the monarchy and
causing tension between
Britain and France

*c.***1760**
Neo-classicism becomes fashionable in
architecture and interior design; Robert Adam
is one of its greatest exponents

1782
James Watt's invention of the rotary
steam engine helps power the
Industrial Revolution

ABOVE: This decorative oval plaque depicting a classical female figure adorns an ebonised cabinet in the drawing room at Llanerchaeron, in Ceredigion, Wales. It is made of Jasper ware – a fine-grained stoneware manufactured by Wedgwood from 1777.

acquired on the Grand Tour – a cultural exploration of the northern Mediterranean. Meanwhile, society painters such as Thomas Gainsborough picked up handsome commissions by portraying his subjects set against a picturesque landscape which reflected a patron's pride in his country house estate.

As a result of the Acts of Union of 1706 and 1707, England, Wales and Scotland were finally united as Great Britain; what had been an uneasy truce became – eventually – a successful, beneficial partnership. For a variety of practical, climatic and cultural reasons, this small cluster of islands off the coast of northern Europe became the birthplace of the world's first Industrial Revolution. There was a general mood of scientific enquiry, an overriding curiosity about the natural world, and plenty of wealthy backers willing to invest in promising individuals. James Watt, whose Scottish family were too poor to have him educated, developed the steam engine, which transformed the generation of power; the potter Josiah Wedgwood was able to combine his practical knowledge with business acumen and a timely marriage to a prosperous third cousin to produce classically inspired artefacts good enough for the King.

Royal Progresses

British monarchs had a long tradition of conducting 'royal progresses' – peripatetic ambles around the wealthiest of their subjects, accompanied by a substantial retinue of courtiers and servants. The monarchs invited themselves to stay and the noble families sometimes nearly bankrupted themselves to entertain their honoured guests, gratifying every whim. Potential preferment or, equally, social ostracism could result from a royal visit. Queen Elizabeth, never the easiest of guests, made disparaging remarks one evening about the size of the forecourt at Osterley Park, in Middlesex, so while she slept Sir Thomas Gresham had a dividing brick wall built across the space to humour her.

Between 1702 and 1712, Queen Anne (1665–1714), the last of the Stuart monarchs, made a number of royal progresses from London to Bath to 'take the waters'. *En route*, she and her slow-moving entourage were accommodated in fine style at the carefully selected homes of the very affluent. Avebury lay on that route, and the picturesque village with its curious standing stones had become a landmark, a curiosity investigated and illustrated by the dogged antiquarian William Stukeley. In 1694 the wealthy lawyer Sir Richard Holford had bought Avebury Manor from the Stawell family for £7,500 and may have redecorated it in the hope that Queen Anne might be his guest. There is no reliable evidence that the monarch ever stayed in the Queen Anne Bedchamber on the first floor. However, in an intriguing manuscript written in 1712 by John Sanders – a servant of Lady Holford's sister – entitled *The Account of My Travils with my Mistress*, the writer claimed:

'Thursday ye 13 [August 1712] about 10 we came to Sir Rich. Holford's house in Avebery, it is a noble larg ancient seat, built with whit larg stone, it did belong to Lord Stowel, ye late noble Lord Stowel was born thare, and our Queen Anne dined thare.'

ABOVE: This painting, after Sir Godfrey Kneller, depicts Queen Anne with the Duke of Gloucester, the only one of her seventeen children to survive infancy.

TAKING THE WATERS

Thanks to its natural hot springs, Bath became a spa in Roman times and was much favoured by the colonisers. The supposed health-giving properties of the baths maintained their reputation into the Middle Ages and, although baths generally became known as places of ill-repute, by the Elizabethan period Bath was increasingly seen as a good place to try a 'cure', a remedy against bodily ills. Queen Anne, who was obese, and suffered from gout along with some severe gynaecological problems, is known to have visited the spa at least five times. Royal patronage turned the city into a fashionable resort and 'taking the waters' remained popular well into the era of Jane Austen.

ABOVE: The elaborate State Bed at Dyrham Park was made in *c*.1704 in the fashionable Anglo-Dutch style. Draped in sumptuous crimson and yellow velvet, with a sprigged satin interior, the bed still stands in the new state apartments Blathwayt built along Dyrham's east front.

During this era, 'dining' took place during daylight hours, so it is possible that Queen Anne, never one to skip a meal, may have broken her journey at Avebury Manor, lunched and travelled onwards. In the ensuing years, however, the tradition developed that Anne slept at Avebury rather than just dined here.

Enticing a monarch

Many wealthy individuals were keen to attract the patronage of the reigning monarch; Dyrham Park, Gloucestershire, was a newly built house which had been furnished from top to bottom with art and artefacts to reflect the status of its owner, William Blathwayt. He installed a State Bed to attract Queen Anne; sadly she never visited. At Osterley, as late as the 1760s, Robert Adam was commissioned by the Child family to create a linked suite of

It is possible that Queen Anne may have broken her journey at Avebury Manor, lunched and travelled onwards.

state apartments, including a bed of almost theatrical opulence; again, no monarch ever slumbered under its canopy, but the family had demonstrated that they moved in the circles which *might* have to accommodate royalty at any moment. State apartments were status symbols – they were not expected to be used.

After Anne's death the Crown passed to the Elector of Hanover (1660–1727), who became George I. He preferred to divide his time between the Hanoverian and English courts and his immediate circle of acquaintances. His son, George II (1683–1760), had a similar

horror of the provinces. However, George III (1738–1820) and his wife, Queen Charlotte, accompanied by friends and family, revived the idea of the 'royal visit' and made themselves more popular by it, staying at well-appointed houses such as Saltram, near Plymouth, in order to show themselves to their subjects.

ABOVE: At Osterley Park in Middlesex, Robert Adam created a suitably opulent plasterwork ceiling for the State Bedroom. The central medallion was inspired by Angelica Kauffman's painting of Aglaia, one of the Three Graces.

RIGHT: Seen here resplendent in full regal robes in a painting by Allan Ramsay, George III revived the practice of the 'royal visit'.

SALTRAM HOUSE

Saltram near Plymouth is one of Britain's best-preserved early Georgian manors. Like Avebury Manor, it was originally a Tudor house which had grown organically from older foundations. Houses change as fashions and fortunes allow, and the catalyst is usually a wealthy and dynamic family. The Parkers bought Saltram in 1712 and commissioned a Neo-classical exterior to mask the irregular structure. Little expense was spared on the interior; the plasterwork and joinery in the main rooms was superb, and at least four rooms were decorated with hand-painted Chinese wallpaper, an exotic and expensive taste. The walls were covered with portraits, many by Joshua Reynolds, a locally born genius who became the leading British painter of his age.

In 1768, the year he inherited Saltram, the younger John Parker engaged the extraordinary architect and designer Robert Adam (1728–92) to transform the interior of his family home. Adam used the very best contemporary artists and craftspeople; every element of his vision was designed or made specifically for this house. In August 1789, the Parkers received the ultimate accolade: George III, Queen Charlotte and their retinue came to stay at Saltram during the royal household's visit to the West Country.

LEFT: The main façade at Saltram has an imposing Doric portico. The central pediment bears the coat of arms of the 1st Earl of Morley, cast in patent Coade stone in 1812.

BELOW: The saloon at Saltram is believed to be one of Robert Adam's finest interiors. This detail shows plasterwork on the coved ceiling.

THE GEORGIANS AT HOME

ABOVE: The *enfilade* at Wimpole Hall, Cambridgeshire, creates a long vista that stretches from the dining room through to the Book Room. The original hall dates from 1640 but was much altered and added to in the eighteenth century.

Within substantial houses, subtle changes were underway both physically and socially after the insecurity of much of the seventeenth century, with its civil war, religious extremism and royal power struggles. Householders still relied on their servants, but now the lower status of

Clients wanted their architects to provide them with corridors with separate bedrooms on one or both sides, each room a sanctuary behind a stout door.

the latter was emphasised. In former eras, bedrooms had led one into another in an *enfilade* format, essentially a line of chambers linked by doors opening from one into the next, with inevitable compromises for personal privacy. By the middle of the eighteenth century, clients wanted their architects to provide them with corridors with separate bedrooms on one or both sides, each room a sanctuary behind a stout door.

In the past, servants had literally 'waited' on their masters, sitting, standing or even sleeping, always within earshot. The innovation of the bell-pull system allowed the family to summon a servant from some distance away, even as far as the basement, where many were now housed for a large part of the day while they tackled mundane chores. 'Above stairs', a new, smarter type of servant had evolved.

THE GRAND TOUR

During the eighteenth century, wealthy young gentlemen gained first-hand experience of the classical world by travelling through France and across the Alps to the northern Mediterranean, often in the company of a tutor who could guide them through the complexities of ancient Roman and Renaissance art. Frequently lasting a year or more, these Grand Tours might take them to Paris, Geneva, Venice, Rome and, later in the century, as far as Sicily and Greece, where they would extend their studies to encompass ancient Greek art. *En route* the travellers would acquire as many precious artefacts as their generous purses would allow, shipping their purchases home in crates for display in their opulent Neo-classical mansions.

Footmen were menservants originally employed to accompany the family on journeys, acting as bodyguards and running errands; now they were dressed in smart uniforms, assisting at table and gracing all social occasions, usually in symmetrical pairs. Fashionable hostesses tried to 'match' pairs of footmen, like twin coach-horses, and recruitment often hinged on a candidate's ability to fit into his predecessor's uniform; livery was expensive. By contrast, other servants, particularly women in service, were expected to provide their own clothing as they worked behind the scenes.

RIGHT: Bell pulls transformed life in large country houses and could even form part of the decorative scheme, as with this ornate example in the Red Drawing Room at Belton House, Lincolnshire.

ABOVE: Palladio's Villa Rotonda (1567), near Vicenza in Italy, inspired English architects in the early eighteenth century and provided the model for Lord Burlington's Chiswick House in Middlesex, completed 1729.

BELOW: This bust of Sir Isaac Newton is one of many in the Temple of British Worthies at Stowe Landscape Gardens, Buckinghamshire, designed by William Kent with sculptures by John Rysbrack and Peter Scheemakers.

The classical style

Stylistically, Palladianism and, later, Neo-classicism dominated architecture, interior design and the decorative arts. Andrea Palladio, who had died in 1580, left consummate designs based on the architecture of classical Italy which greatly appealed to the eighteenth-century Whig aristocracy and meritocracy. As educated men, Avebury owners Sir Richard Holford and Sir Adam Williamson would have been familiar with the visual vocabulary of the Ancient Romans, as well as their writings, philosophy, mythology and even battle tactics. The rationality of the architectural dimensions and their symmetry appealed to orderly minds; there was also a Roman-inspired cult of portraying 'great men' through the work of celebrated sculptors like John Rysbrack, or paintings by men such as Sir Godfrey Kneller, a friend of Sir Richard's.

Architectural idealism was necessarily tempered with realism. In the climate of cold, damp and grey Britain, architects had to strike a pragmatic balance between the 'portico' lifestyle of Italy, with open-air access from one part of the estate to the other, and the need to keep out the winter winds, rain and snow. Grand columned entrances led into porches and lobbies that were designed to trap the incoming cold air and insulate the rest of the building, as at Kedleston Hall, Derbyshire. Architectural motifs such as fluted columns were adapted to support hefty marble fireplaces, considered no less classical for being full of reassuringly smouldering logs; triangular pediments, found in ancient Italy on entrances to Roman temples, were added atop stout, well-fitting oak doors in provincial dining rooms from Edinburgh to Cheltenham.

BELOW: Designed by Robert Adam in the 1760s, the monumental Marble Hall at Kedleston is top-lit to suggest the open courtyard of a Roman villa.

THE NEW CLASSICISM

In the early seventeenth century, Inigo Jones transformed English architecture by introducing a classical style based on the works of Andrea Palladio. Buildings such as his Banqueting House in Whitehall, London (1619–22), were to inspire architects for generations to come. Following the Restoration in 1660, however, Jones's influence waned and country house design came under the influence of Dutch models. The rectangular 'double-pile' house became fashionable, with its tall hipped roof pierced with dormer windows, and projecting centrepiece topped with a pediment.

A brief flowering of the English Baroque in country house design around the turn of the century resulted in such masterpieces as Blenheim Palace, in Oxfordshire (1705–24). But with the accession to the throne of George I and the return to power of the Whig aristocracy, such florid expressions of wealth were considered to be too foreign and Catholic in taste, and a new style of classical architecture emerged. Based on the designs of Inigo Jones and Palladio, this was to prove hugely influential over the next forty years and encouraged house owners such as those at Avebury Manor to remodel their homes in the leading fashion of their day.

ABOVE: Built in 1690, Uppark, in West Sussex, was built in the then-fashionable Dutch style, with a tall hipped roof, dormer windows and a pediment over the three central bays.

Palladianism

The founder of this new school of architecture was Scottish architect Colen Campbell, whose highly influential book, *Vitruvius Britannicus* (published in three volumes between 1715 and 1725), depicted the work of English architects from Jones onwards, including a number of the author's own designs. Key practitioners of Palladianism were Lord Burlington, whose Chiswick House was based on Palladio's Villa Rotonda, and architects William Kent and James Paine. Buildings such as Kent's Holkham Hall, Norfolk (begun in 1734) and Paine's Kedleston Hall, in Derbyshire (begun 1757), typified the style, with a grand portico based on a Roman temple front, flat roofline, tall *piano nobile* (principal floor), and internal architectural elements including classical columns and pediments. These features would filter down the architectural hierarchy to influence the design of smaller houses and terraces across the country.

The Neo-classical style

By the 1750s, however, architects were increasingly turning to Antique models for inspiration, travelling to Rome, Greece and beyond to study the ancient sites first hand. Here they discovered a vast source of architectural vocabulary, displayed on buildings as diverse as temples, triumphal arches and public baths, which they reinterpreted and used to enrich their own work. One of the greatest exponents of this new style was Scottish architect Robert Adam, who introduced Antique references into his exquisitely designed interiors, covering walls and ceiling with elegant arabesques, swags and medallions in paint or plaster.

BELOW: Robert Adam's elegant form of Neo-classicism reached a pinnacle of perfection in the Etruscan Dressing Room at Osterley Park, London which Adam designed in the 1770s.

ABOVE: With its six giant Corinthian columns and triangular pediment, the grand portico at Kedleston Hall, Derbyshire was designed to resemble a Roman temple front.

Walls, floors and windows

Magnificent stone halls laid with intricately patterned marble floors might impress visitors but, for all their chilly grandeur, they lacked warmth. What most householders wanted underfoot for at least six months of the year was an opulent, handmade woollen carpet. Weavers at Wilton, in Wiltshire, first built a carpet factory in 1655 but it was not until the mid eighteenth century that technological innovations enabled England to produce large scale carpets to rival those on the Continent. Ideally the carpet pattern should be made to

ABOVE: A mid-eighteenth-century Rococo pier-glass, set over a gilt console table in the drawing room at Ickworth House, Suffolk, would once have reflected the flickering candlelight.

BELOW: In the saloon, or Great Drawing Room, at Saltram, designed by Robert Adam in 1768, the pattern of the Axminster carpet echoes that of the plasterwork ceiling.

echo the design of the moulded plaster ceiling just completed by some authentically Latin artisans engaged by one's architect. Opulent fabrics such as hefty draught-proof damask curtains were also much in demand, as were light-radiating large-scale mirrors, set in carved gilt frames.

Wall-coverings were in many respects the easiest way of transforming an interior and in the early part of the century they were usually made of narrow bands of woven damask fabric, seamed together and fixed onto battens. However, by the end of the century it was more common for noble rooms to have a painted finish, with architectural details highlighted in gilt, or even a printed wallpaper.

ABOVE: '*Les Deux Pigeons*' wallpaper in the Palladio Room at Clandon Park in Surrey was made *c.*1780 by Jean-Baptiste Réveillon, the leading French wallpaper maker of his day.

Georgian furniture

Throughout the Georgian period furniture gradually became lighter and more sophisticated, more portable and decorative, and better suited to the Age of Elegance. Neat chests of drawers were a great improvement on the cavernous trunks of old, while folding pieces such as games tables were adaptable and well-suited to life in tall, narrow city houses.

Where possible in the eighteenth-century interior, individual pieces of furniture were placed against the wall, in symmetrical arrangements of straight-backed chairs, sofas and side-tables; these pieces could be moved into the room and grouped as required. Dining rooms were similarly unencumbered; trestle tables with extra leaves would be stored in an anteroom then brought in to make up a table of any size required, topped with a linen cloth, a dinner service of fine china, hand-made glass, and ample silver.

ABOVE: A French bombé commode, with its swollen shape and ormolu mounts, was the height of fashion in the mid-eighteenth century. This one stands in the Large Drawing Room of The Vyne, in Hampshire.

The oriental vogue

While British interior design was predominantly classical throughout the eighteenth century, decoratively it also reflected the prevailing passion for objects and artefacts from Asia and the Far East, especially ceramics, lacquerware, textiles and other goods from China. Porcelain was a particular obsession; collectors wanted to acquire blue-and-white plates, vases and bowls, as well as *famille rose* or *verte* (multicoloured with predominantly pink or green schemes), ideally buying pieces in pairs to fit their symmetrical classical interiors.

The taste for *chinoiserie* also influenced indigenous British makers; skilled furniture-makers such as Thomas Chippendale met the desire for Chinese-style dining chairs and sideboards, while cabinet-makers used imported lacquered panels in their 'japanned' commodes and cupboards.

The craze fed the demand among the wealthy for 'Chinese' rooms fitted with hand-made wallpapers custom-made in Canton (now known as Guangzhou). Supplied with the correct measurements, and using skills from a long tradition of scroll painting, specialist Chinese merchants could supply entire room-schemes of bespoke wallpapers, painted by hand, to be installed in manor houses on the other side of the world.

RIGHT: A pair of Chinese *famille-rose* porcelain vases (*c.*1760) adorns the window recesses at Polesden Lacey in Surrey. *Famille rose* was distinguished by its rose pink pigment, which was probably introduced from Europe where it was called purple of Cassius.

THOMAS CHIPPENDALE

The cabinet-maker Thomas Chippendale (1718–79) moved from his native Yorkshire to London around 1745. He is best-known for his influential book of designs, *The Gentleman's and Cabinet Maker's Directory* of 1754. So popular and so widely imitated were his designs that it can be difficult to definitively identify work produced from his premises. Thomas Chippendale had a succession of partners and his son, also called Thomas, took over the business after his death in 1779. Throughout his working life Chippendale Senior was adept at keeping abreast with fashion, encompassing the taste for Rococo and Chinese styles (see right) as well as making furniture designed by the Adam brothers.

A Century of Change

In 1694, Sir Richard Holford bought Avebury Manor for £7500 from Lord Stawell. He was 65 years old, a successful barrister and Master in Chancery, and came from a family of London-based property speculators. His first marriage had brought him Westonbirt, an estate in Gloucestershire; his second gave him heirs. On his third marriage, to a Miss Suzanna Trotman in 1689, a pre-nuptial contract stipulated that he must provide another house for her because of his comparatively advanced years and existing adult children.

Sir Richard and Lady Suzanna were considered enlightened individuals by the standards of the age. He was an active landlord, repairing stonework, building walls and improving the Manor's gardens. In keeping with the tradition of the times, he may also have decorated the manor in preparation for a visit from Queen Anne.

Suzanna's preoccupations were different; she left £200 for the education of poor children in Avebury in her will. The first schoolmaster

ABOVE: The first Sir Richard Holford, seen here in a portrait by George Vertue, bought Avebury Manor as a home for his third wife, Suzanna Trotman.

BELOW: One of Richard Holford's direct descendants established the now world-famous Westonbirt Arboretum in the mid-nineteenth century.

was appointed in 1736 to teach children in the church and by 1783 the number of pupils had grown to sixteen. The school was eventually superseded by the National School and the charitable funds left by Suzanna continued to help support it.

Sir Richard died in 1718, leaving the house and estate to Suzanna. Their son Samuel, the only child of seven to survive to adulthood, inherited in 1722, but in 1728 he died and the property reverted to the grandson of Sir Richard's first marriage, another Richard Holford. It was during the second Sir Richard's ownership that Avebury was remodelled in the early Georgian style; the Great Hall was converted into a grand dining room, with carved and pedimented doorways and a fine mantelpiece, while the State Bedchamber gained a deep plaster cove and ornate ceiling. On his death in 1742, Avebury Manor was occupied by Richard's brother Staynor Holford, who lived there with his mother and half-brother Arthur Jones.

The Joneses take over

When Staynor died in 1767 Arthur Jones inherited Avebury Manor, and the responsibility obsessed him for the next 22 years; he constantly changed his will, finally settling on his niece, Ann, known as 'Nanny', who was married to Adam Williamson, a military man. She visited him in 1789 and recorded his meticulous instructions that in the event of serious illness his feet, stomach and bedding should be kept warm. He also insisted that after his death his coffin should not be closed nor buried for a week – the terror of accidental interment was a common preoccupation of the age.

After 'Nanny' Williamson inherited Avebury Manor she lived there for a period before perishing of yellow fever in the West Indies, where she had gone to join her husband. Sir Adam returned alone to Avebury and on his sudden death in 1798 the property was inherited by Richard Jones, nephew of Arthur Jones. In just a little over a century Avebury Manor had come under the ownership of nine individuals, linked through the intricate ties of blood and marriage.

ABOVE: Magnificent pediments in the early Georgian style were placed over the doors when the dining room was remodelled in the 1740s.

THE QUEEN ANNE
BEDCHAMBER

ocal tradition holds that Queen Anne stayed at Avebury Manor while travelling by coach from London to Bath to take the waters but, although the Great Chamber has long been named after her, this story has never been verified. There is, however, documentary evidence that she dined at the house at least once, and it is therefore likely that Sir Richard Holford redecorated the 'Queen Anne Bedchamber' in anticipation of the monarch choosing to break her journey at his home.

BELOW: This photograph shows the Queen Anne Bedchamber as it was in the 1920s, with its deep-coved ceiling and bed hangings embroidered by Nora Jenner.

During a royal progress, in an era before reliable domestic plumbing, a courtier or personal attendant would discreetly hint to the honoured householder that the visiting monarch wished to 'retire' or 'rest' in privacy. Such euphemisms covered that universal human need: even a monarch needed to use the lavatory! At Avebury, the preparation of a grand suite of rooms – a formal bedroom with anterooms attached, all freshly decorated in the latest fashion and equipped with a discreet close stool – would have been very welcome. For Queen Anne, as for later monarchs, perhaps there would have been a vague sense of puzzlement that everywhere she visited smelt of fresh paint.

The mystery behind the history

The Queen Anne Bedchamber at Avebury Manor sits above the dining room in the south wing, the first and largest room in a suite of rooms that run one into another, a typical arrangement for this era. However, if the queen did indeed stay here, there are some stylistic anomalies. The room has a distinctive deep-coved

ABOVE: The plasterwork on the ceiling includes a Greek key pattern, introduced into decorative schemes in England in the early Georgian period.

LEFT: State Apartments, such as those at Kedleston Hall in Derbyshire, were designed with one room leading into another. Passing through them was like taking a journey from the most formal room, the ceremonial bedchamber, to the most intimate – the withdrawing room.

ceiling reaching into the roof space which architectural experts consider to be a later addition, dating from some time between the 1720s and the 1740s. If this is the case, the room was remodelled after the deaths of Queen Anne (in 1714) and her ambitious would-be host, Sir Richard Holford, who died in 1718.

Until recently, documentary evidence to support this theory was scanty, although William Stukeley – that dogged recorder of all things concerning Avebury – left a rough sketch of Avebury Manor positively dated to 1723.

Crucially, his drawing shows a series of gables along the south façade, rather than the parapet we see today. It is unlikely that the tall coved ceiling could have been built while the timber structure of the gables was in place, which means the room was remodelled some time after 1723.

Solving the mystery

The mystery was finally resolved by the findings of the Wessex Archaeology Survey (WAS). Tree-ring dating demonstrated that roof timbers in this part of the house were felled in about 1600, corresponding with the date of construction of the south wing. But there was also physical evidence that the roof had been modified at a later date; the lower tie beams of the trusses over the bedroom had been removed and their function replaced by a 'collar' at a higher level, thus providing space for a new deep-coved ceiling.

The question was: who had done the work, and when? Luckily, Wessex Archaeology were able to corroborate their findings. In a local archive they found an invoice, dating from 1740, for delivery to the manor house of large quantities of lead, which could only mean a major reroofing programme was underway. This was convincing proof that the remodelling of Avebury Manor, including the building of a new parapet and the insertion of a coved ceiling in the Queen Anne Bedchamber, took place in 1740, at a time when Richard Holford's grandson, another Richard, was owner of the manor.

ABOVE: The Queen Anne Bedchamber during the 1970s, when Lord Ailesbury owned Avebury Manor.

TRANSFORMING THE BEDCHAMBER

W hen Russell Sage and the experts considered their design approach to this room, the first thing they had to decide was: to which 'moment' should the room be returned? Throughout this project they had concentrated on re-creating the heyday of each interior, the moment when a particular personality, or a historical event, was most prominent. So, accepting Queen Anne's associations with Avebury, they decided the room should be decorated and furnished as though peripatetic royalty was expected at any moment. For the ceiling, although of a later date, an appropriate decorative scheme was devised, based on contemporary interiors from the early 1700s, while the painted decorations on the walls were designed to be impressive and dignified. Meanwhile the room would be filled with sumptuous textiles and furniture fit for a queen and specially commissioned to reflect the chosen period.

There are strong precedents for designing this room to suit an earlier stage in its history. In 1910, for example, the Jenners commissioned a Mr Titcombe to make a frame for a four-poster bed in the Queen Anne style, for which Nora Jenner skilfully embroidered bed hangings and coverings.

BELOW: During the presentation day, designer Russell Sage drapes rich fabrics around the empty room to give an impression of how it will look with the bed hangings in place.

ABOVE: Documentary evidence on the Balcony Room at Dyrham Park provided inspiration for the marbling.

BELOW: Expert Dan Cruickshank comments on the decorative scheme on the walls of the bedchamber.

A regal decor

Russell and the experts decided upon a scheme of paint effects to represent marble, based on their observation of an old marbling or wood grained effect still visible on the wooden window shutters in the adjacent Withdrawing Room. They were inspired by a number of early schemes – both existing and documented – including the Painted Room (of *c.*1700) at Hill Court near Ross-on-Wye, the Balcony Room of the mid 1690s at Dyrham Park, Gloucestershire, and the 1670s interior of Ham House, Surrey. Expert Dan Cruickshank also recommended the dramatic *trompe l'oeil* effects found at Het Loo Palace at Apeldoorn in the Netherlands, completed in 1686 for the future William III of England, and with which Sir Richard Holford would have been familiar.

At Avebury Manor, versatile fine decorator Grant Watt and specialist painter Corin Sands used their restoration experience to create a complex paint scheme on the walls, in homage to the finest marbling techniques

MARBLING AND GILDING

Around 1700, graining and marbling were frequently used for decoration in houses of all classes; skilled painters could depict the subtle veining and textures of all sorts of fine materials, from serpentine to cedarwood, lapus lazuli to malachite. In the Queen's Closet at Ham House (right), the marbled plasterwork on the ceiling is combined with gilding, which would have added a subtle touch of richness when seen by flickering candlelight. A document from 1694, concerning the décor of the Balcony Room at Dyrham Park, near Bath, states that the paint effect on the panelling was originally a dark red porphyry colour combined with orange-pink marbling (presumably in imitation of Sienna marble). This description provided the inspiration for the decorative scheme of the Queen Anne Bedchamber at Avebury.

available in the early eighteenth century. Varnish was applied to a white ground to keep the subsequent layers of paint receptive to manipulation and prevent them from drying out; marbling requires the careful application of several coats of contrasting colours, using different tools at each stage, to achieve a subtle, rich texture and vibrancy of colour. The tools used are variously sponges, brushes and feathers.

The coving, meanwhile, was painted to look like clouds. Grand ceilings of this era were often decorated with *trompe l'oeil* skies and clouds, as in the Green Closet at Ham House, implying a heavenly vista beyond. Corin and

RIGHT: Expert Anna Whitelock admires the finished decorative scheme in the Queen Anne Bedchamber, which includes marbling, sponging and a *trompe l'oeil* sky effect on the coving.

his brother Ashley added touches of gilding to the cornicing for a sense of regal splendour. All the paints, dry pigments and clear glazes employed to build up subtle effects were supplied by Earthborn Paints, whose products are manufactured without the use of acrylics or oils. This means they are very porous, allowing the wall surface to breathe, a necessity in many old buildings. The skirting and raised mouldings, in a contrasting off-white marble, were painted last to avoid splashes from above.

Making the State Bed

Russell commissioned specialist furniture restorer Dave Lyons to make the frame of a replica State Bed in the Queen Anne style, incorporating some old bed posts bought at auction. On Dan's advice, two pairs of posts were selected. Although these dated from the Regency period, their proportions were good and the nineteenth-century details would be obscured with fabric. Much thought and deliberation went into creating the bed rails and the frame around the top of the posts to achieve the correct size and proportions. The magnificent State Bed at Dyrham Park was a key inspiration, though examples in the Victoria and Albert Museum and Dan's considerable expertise contributed to the final design.

A flat tester would not have suited the coved ceiling in this room and Russell wanted the bed to have a domed top, almost architectural in inspiration. A blacksmith made the supporting structure for the dome, which was then swathed in fabric. The final challenge was to create a complex moulded cornice in a number of detachable sections to go around the top of the bed, on the sides and across the front. This key component of the bed was achieved with the help of furniture designer James Howett and joiner Aaron McGill who is experienced in producing traditional details. The various mouldings and shaped blocks making up the ornate cornice are covered with fabric, so they have to be detachable from the back yet conceal the arrangements for hanging the drapes and curtains.

ABOVE: The delicate veining on the orange-pink marbled walls is carefully applied using a fine paintbrush.

RIGHT: Presenter Penelope Keith visits Dyrham Park, Gloucestershire, to inspect the State Bed, on loan from the Lady Lever Art Gallery. The original red and gold bed hangings have been removed for restoration.

SLEEPING IN STATE

During the seventeenth and eighteenth centuries, having a State Bedroom in one's own home was an essential for the grandest society figures because it was an indication that royalty might be coming to stay. These bedrooms formed part of the State Apartments, a suite of rooms which tended to be created by the most notable architects and craftsmen of the day; Robert Adam, for example, provided a State Bedroom at Kedleston Hall, Derbyshire, for Lord Scarsdale in 1758, and a suite of State rooms for the Child family at Osterley Park, Middlesex, in 1776.

Ironically, State Bedrooms were rarely used for sleeping in; once formally prepared for sleep by attendants, the monarch would retire to a more private anteroom for the night. He or she would then return to the State Bed in the morning for the *levée*, the public business of going through one's toilette and receiving honoured guests.

BELOW: The highly elaborate State Bedchamber at Osterley Park, Middlesex, was designed by Robert Adam.

Perhaps because they were rarely used, many State Beds still exist in historic houses, though complete State Bedrooms are less commonly found, the rooms having been restyled for other purposes. However, the State Bedroom at Powis Castle, Powys, dating from the 1660s, survives intact and is the only one in Britain to have a balustrade railing off the bed alcove from the rest of the room.

For all its sumptuous crimson and yellow velvet, the splendid State Bed at Dyrham Park in Gloucestershire, dating from 1704, was never slept in by royalty. Only slightly less opulent is the crimson and gold State Bed sent from Windsor Castle to Warwick Castle, Warwickshire, in advance of a visit planned by Queen Anne in 1704 which was subsequently cancelled. Sending one's bed in advance was a royal fixation. The State Bedchamber at Wilton in Wiltshire had no bed, so when George III and Queen Charlotte planned a visit, a suitably ornate example was transported from neighbouring Fonthill Abbey. Taking no chances, the royal household also dispatched a bed, and the two were set up in the same room.

A State Bed was a four-poster with an ornate canopy and hanging curtains. A rectangular base of canvas was laced tightly to the bedframe, then topped with a straw palliasse and three or four wool-lined mattresses. Feather mattresses added the final layer to the snug pile, and numerous pillows and bolsters were provided to prop up the sleeper. Such a high structure required a set of wooden steps for access, which often contained cupboards to conceal chamberpots.

RIGHT: The State Bed at Calke Abbey, Derbyshire, was probably made for George I *c.*1715. It has Chinese embroidered silk hangings and is in immaculate condition.

ABOVE: The ornate State Bedroom at Powis Castle, Powys, dates from the 1660s. The original furnishings include the State Bed, which sits in an alcove, separated from the rest of the room by a balustrade rail.

The bed hangings and textiles

For the opulent fabrics to dress the bed and provide the curtains, Gainsborough Silk Weaving Company was commissioned to produce 82m (90yd) of red silk damask. Silk has always been a favourite of the wealthy, being strong, versatile and able to take a variety of dyestuffs to create richly coloured fabrics. It is very appropriate for this setting as the silk-weaving industry took root in Britain towards the end of the seventeenth century with the influx of Huguenots from France, who brought their expertise in silk-weaving with them.

ABOVE: Gainsborough employs a CAD system in their design process.

BELOW: The weft yarn is mounted into the shuttle on one of the 100-year-old Hattersley Jacquard looms.

Gainsborough specialises in creating high-quality historic furnishing fabrics and they undertook every aspect of the production in-house, consulting their archives of more than 2,000 historic designs for approval by the design team. They used traditional techniques to prepare and dye the yarns and wove the textiles on a turn-of-the-century Hattersley Jacquard loom.

The Avebury Tassel

Russell commissioned a set of large, ornate, unique pieces to 'crown' the top of bedposts, above the decorative bed cornice. Each 'Avebury Tassel' comprises a carefully turned wooden base, reminiscent of an urn, which is then covered in luxurious silks, dyed in colours to match the grandiose setting. These were made by the firm Henry Newbery, first established in 1782, which specialises in designing and making intricate, authentic trimmings for upholstery, curtains and interiors, known collectively as *passementerie*. Every aspect of their furnishing trimmings for this bedchamber, the State Bed and the adjacent furniture is appropriate to the early eighteenth century, referring to historical examples as well as their own archives and collections.

The Bedroom Furniture

The firm ELG at Sainsburys, specialists in supplying hand-carved reproduction furniture, rose to the challenge of making new furniture in the Queen Anne style for the State Bedchamber. The brief was to produce appropriate furniture inspired by historical examples – 're-creations' rather than 'replicas' – and as always the design process was a collaborative one. Thomas Sainsbury photographed pieces of furniture *in situ* at Dyrham Park, dating from the first decade of the eighteenth century, and measured them minutely. A life-size drawing of each was made, to be copied by the cabinet-makers, rather like using a paper dress pattern. Blocks were made with their exact joints already cut out, turned in-house then sent to specialist carvers where necessary. The finished components were assembled like a 3-D jigsaw.

ABOVE: Gainsborough have their own custom-built dye house, which enables them to hand dye all their yarns and achieve the exact colours they require.

The firm has made six upholstered high-backed chairs (four for the State Bedchamber, two for the adjoining anteroom), each with carved front legs and a carved stretcher; the original had arms, but the modern interpretation has been made without them. The chairs are stuffed with horsehair, painstakingly upholstered with fabric to match the bed hangings, and finished with braid. There is a co-ordinating stool at the foot of the bed, also based on an example at Dyrham. Across the room is a two-seater sofa, with turned legs, upholstered arms and a high 'camelback' (double humped), based on the individual chairs.

An upright pier mirror, constructed from hardwood and entirely bench made, has aged glass and a double arched top to the moulded frame, reflecting the shape of the chairs and the sofa back; the carved crest is typical of the style associated with George II. The water gilding was created by traditional methods, using 22 carat gold leaf. All the State Bedchamber furniture has been made in solid walnut, bleached out to achieve the golden patina normally endowed by centuries of gentle sunshine.

BELOW: Cabinet-maker Jason Corbett at ELG carries out the final assembly work on one of the Queen Anne chairs.

The Finishing Touches

To add a finishing touch, Thomasina Smith was commissioned to create a portrait of Queen Anne to hang in the Bedchamber. Describing herself as an art historian, artist and artisan, Thomasina creates paintings in historical styles and her work has appeared in many films, including *Notting Hill* and the *Harry Potter* series. For this room she painted an unusual depiction of the Queen, drawing on the triple portrait of Charles I by Van Dyck, painted in 1635 and now in Windsor Castle; multiple portraits were not uncommon in the seventeenth and eighteenth centuries, when they were often executed as a preparation for a sculpture.

For research purposes, Thomasina examined a number of contemporary portraits, including a statue in Queen Anne's Gate in central London. The picture depicts the head and shoulders of the monarch in three different poses, and represents how she would have looked in about 1705, aged around 40, wearing a coronation crown, with pearls and lace at her neck. The painting has been executed in light and fresh colours in oils on canvas, and measures 1.5m × 0.5m (5ft × 1ft 7in) to fit the niche over the fireplace when surrounded by a plaster frame.

BELOW: Van Dyke's famous triple portrait of Charles I was painted in preparation for a sculpture, to be made by Gianlorenzo Bernini.

BELOW: Artist Thomasina Smith's triple portrait of Queen Anne was inspired by the portrait of Charles I and is based on several contemporary portraits of the queen.

OVERLEAF: The finished Queen Anne Bedchamber is resplendent with its period style bed, lavish silk hangings by Gainsborough, and bespoke furniture by ELG at Sainsburys.

Transforming the Antechamber

Leading from the Bedchamber, and forming part of the Queen Anne State Apartments, is a discreet antechamber, lit by a small window, which was traditionally called the Chinese Print Room. In earlier eras, an antechamber acted variously as a dressing room, a cosy boudoir, or a quiet waiting room where people gathered prior to admission to the Withdrawing Room beyond – the 'inner sanctum' where royalty resided. Those privileged enough to approach the royal presence might wait here, as did body-servants to the monarch, ready to take their instructions. An antechamber also formed a physical barrier between the bedchamber and the closet, a small adjacent room containing the close stool. Both Ham House and Dyrham Park still have similar antechambers – small, transitional places, with few pieces of furniture, but those of the highest quality.

A black and white photograph from the time of Alexander Keiller's occupation in the 1930s shows that this room was decorated in the early eighteenth-century Chinese taste, with an authentic Chinese wallpaper (possibly mounted on panels lining the room), some Chinese-style paintings on mirrors, and a few ornate lacquered cabinets, a style very much associated with aristocratic and royal settings. 'Chinese' rooms of this type were expensive status symbols. By the 1690s it was possible to buy from London dealers individual rolls of painted paper, known erroneously as 'Indian' papers, because they were mostly

Above: The Queen's Antechamber at Ham House, Surrey, is decorated with silk velvet wall hangings and contains high quality pieces of furniture and artefacts.

imported from China by the East India Company. Typically, they depicted garden scenes, hand-painted on a common-coloured ground and ornamented with exotic birds and flowers, and roll after roll could be hung alongside each other in a room to decorative effect. However, such papers were expensive and could be difficult to procure, so enterprising London paper-stainers (the contemporary name for wallpaper-makers) began to exploit this fashion by producing printed papers in imitation of the Chinese imports.

ABOVE: An old photograph from the 1930s shows the exquisitely decorated Antechamber, with its Chinese narrative wallpaper and ornate Oriental-style lacquered cabinet and other furniture.

ABOVE: Specialist painter Mark Sands uses a very fine brush to paint the details onto one of the flowers, supporting his hand on a mahlstick.

BELOW: A length of Chinese printed wallpaper found in the Victoria and Albert Museum provided the inspiration for the wall painting in the Antechamber.

The Design Scheme

As a waiting room for very important people, the Antechamber needed a subtle but rich form of decoration appropriate to the era. Inspiration was found in the Victoria and Albert Museum, where there is a sample of British-made 'Chinese paper', a woodblock print on paper to which was added colour stencilling and a coat of varnish. Made around 1700, it was salvaged from Ord House in Northumberland and has a pattern derived from oriental motifs such as blossoms, squirrels, peacocks, parakeets and a Chinese figure. The design has a repeating pattern, which was not to be found in Chinese wallpapers, but provided a repetitive visual structure more pleasing to Western eyes when applied to British walls.

The colour scheme of this paper is unusual – a dramatic black background, apparently painted freehand, with more pastel figures in the foreground. The design recalls the texture and finish of black lacquered furniture, an effect heightened by the application of varnish – Japanese lacquer was also fashionable in British interiors at this time. A decorative scheme with a similar effect survives in the Lacquer Drawing Room at Burton Agnes in Yorkshire; here the dramatic colour scheme and sinuous, Chinese-inspired decorations were put up in 1715, and the room also houses patriotic portraits of both Queen Anne and her successor, George I, an indication that *chinoiserie* was considered a fitting backdrop for the grandest in the land.

Specialist painter Mark Sands was given the task of creating a historically accurate scheme for this room, setting it in the early eighteenth century. Sir Richard Holford, who owned Avebury at this time, was a habitué of London artistic circles, and would have been very aware of fashionable schemes for interior decoration. He might well have been prepared to commission expensive decorative schemes in order to entice a visit from Queen Anne.

After carrying out extensive research at the Victoria and Albert Museum, Mark Sands hand-painted a Chinese-style scheme with a dark background directly onto the walls, using a large-scale stencil to lay out the pattern. Notwithstanding the oriental style and colour palette, he has included witty and subtle references to the Wiltshire countryside, including flowers that thrive locally, such as the wild pansies so popular in Victorian gardens. Mark also represented indigenous wildlife: red admiral and peacock butterflies, great crested newts and red squirrels; a last-minute addition was a very occidental-looking fox. The whole scheme is a witty and subtle visual pun on the use of Chinese vocabulary and a traditional oriental colour palette to portray a very British fondness for one's own native turf.

BELOW: The lavishly painted walls, with their black background and vibrant colours, give an air of exotic luxury to this small room.

THE ORIENTAL TASTE

Ataste for Chinese objects and artefacts developed in Britain in the late seventeenth century and has undergone periodic resurgences ever since. In the first half of the seventeenth century European trade with China was limited, so in response to the growing popularity of oriental goods Western manufacturers and craftsmen looked for ways to imitate the most marketable products of the Far East. This decorative style of applying oriental motifs and materials to objects and interiors created in the West became known as *chinoiserie*.

The appeal of *chinoiserie* lies in its sophisticated associations with the exotic and picturesque and the fashion for it reached its peak in the mid-eighteenth century, though it remained a recurrent theme in British tastes as late as the 1930s. In modish circles it spread to all forms of the decorative arts, encompassing silver, porcelain and textiles as well as architecture and interiors such as the Chinese rooms at Claydon House.

The visual vocabulary of Chinese style was disseminated by a number of influential publications, including *The Gentleman's and Cabinet-Maker's Directory*, an informative pattern book of furniture designs published by Thomas Chippendale in 1754, and *The New Book of Chinese Designs* by George Edwards and Matthew Darley, which came out the same year and was broader in scope. Three years later, *Designs for Chinese Buildings* by Sir William Chambers introduced the Chinese style as a suitable medium for follies and pavilions, such as his own Chinese Pagoda in Kew Gardens, built in 1762.

For the first time, Western craftsmen could imitate and reinterpret the visual devices found in books, making 'Chinese' fireplaces, doors, cornices, lanterns and mirrors. Certain elements of Chinese aesthetics were also imitated; the asymmetry of pattern and the use of empty space in figurative backgrounds such as wallpaper and textiles was a novelty to designers used to covering every inch of a surface with regularly repeating patterns. The Chinese style offered an exotic, even a witty counterpoint to the more formal Neo-classical styles currently prevailing.

ABOVE: This intricate doorcase in the Chinese Room at Claydon House, Buckinghamshire, was created by the highly talented carver Luke Lightfoot.

It is interesting to note that the shorthand vocabulary of motifs that are still most closely associated with China evolved around this time; dragons chasing pearls, flying herons and mandarin ducks, languid ladies in idyllic gardens, etiolated bearded scholars, intricate pagodas, prunus blossom and lotus flowers, sketchy idealised landscapes, fretted lattices, decorative finials. Even the Chinese colour palette influenced Western tastes, creating a fashion for strong 'Imperial' yellow, lacquer red, and vivid green to adorn the interiors of affluent households.

BELOW: The red japanned bureau-bookcase in the State Bedroom at Erddig in Wrexham, was made in England and dates from the 1720s.

ABOVE: The pier glass and commode in the State Bedchamber at Nostell Priory, Yorkshire, is in the Chinese style and was designed by cabinet-maker Thomas Chippendale.

Transforming the Withdrawing Room

Traditionally at the heart of the State Apartments lay the Withdrawing Room, the inner sanctum, where fortunate commoners might be summoned to meet royalty. Here, rather than in the grandiose State Bedchamber, which was provided more for show than practical purposes, the monarch could sit, sleep, eat or receive guests in privacy.

The Withdrawing Room at Avebury Manor, once known as the Cavalier Room, lies directly off the Antechamber and is believed by some to be haunted. This room dates from around 1600 and lies at the west end of the south wing. Since its transformation it has an air of refined simplicity, with its substantial fireplace and minimal furniture, combined with opulent fabrics suitable for a queen. The robust classical

ABOVE: This delightful heart-shaped window catch dates from the late seventeenth or early eighteenth century.

RIGHT: The Withdrawing Room at Ham House, Surrey, is elegantly covered in crimson damask and displays some pieces of fine Kangxi porcelain.

APPARITIONS AT AVEBURY

Stories persist about the melancholy atmosphere in the panelled Jacobean Crimson Room, or Cavalier Bedroom, at Avebury Manor. There have been numerous sightings of a man wearing a faded red outfit typical of the Cavaliers standing near the window overlooking the south lawn, or to the left of the fireplace, and the figure has also been seen at the window from the garden below. Tradition has it that this is the ghost of Sir John Stawell, a dedicated Royalist, who bought Avebury Manor in 1640. He fought for two years in the English Civil War, was imprisoned in the Tower and his estates were confiscated in 1646. On the Restoration of Charles II in 1660 the Avebury estate was returned to Sir John, but he died only two years later.

When Sir Frances Knowles, a respected scientist, first opened the house to the public in the middle of the twentieth century, visitors would ask him if this room was haunted. The Cavalier's appearance is allegedly preceded by a drop in temperature and the smell of roses, often used in the seventeenth century as an *eau de toilette*. There are also stories of a young woman dressed in white, believed to be Sir John's ward, who committed suicide at the house after the death of her fiancée in the Civil War. Bearded male visitors have reported a peremptory tap on the shoulder, but when they turn round there is no one there.

ABOVE: This detail from a painting at Dunster Castle, Somerset, shows the distinctive collar and long hair worn by Cavaliers. These features are instantly recognizable, whether depicted in a painting or seen as a ghostly apparition.

LEFT: The Cavalier Bedroom, as it was previously called, has provided chilling moments for some visitors to Avebury.

ABOVE: The Closet, seen from the Withdrawing Room prior to the transformation, has some fine oak panelling from *c.*1700, suggesting it was originally a room of high status.

fireplace appears to date from the very early eighteenth century, as do the grained or marbled window shutters. These refined elements seem to confirm that this room was indeed redecorated in expectation of a visit from Queen Anne. The dark wooden wainscotting, now largely hidden beneath fabric panels, is a later addition, probably installed by the Jenners in the early twentieth century.

One discreet door leads into a further small room, the Closet; this may have been used as a small study, perhaps furnished with a desk and a 'cabinet of curiosities', such as fossils, ancient coins or sea-shells. Alternatively, it may have provided a convenient place in which to keep a close stool, as in this case.

Furniture and Fabrics

The day-bed is the focal point of the Withdrawing Room, as this is where the monarch would have rested while receiving guests during a visit. Dan visited ELG at Sainsburys (cabinet-makers) to discuss suitable sources, materials and traditional methods of furniture-making and was particularly impressed by the expertise of Eugene 'Fred' Jennings, a woodturner aged 83, who has worked for the firm for 57 years. The subsequent design process was a team effort, involving Dan, Russell and father and son Jonathan and Thomas Sainsbury, and involved a great deal of consultation of historic sources.

Using models at Dyrham Park for inspiration, the cabinet-makers created a Queen Anne-style day-bed, robust in form and shaped like a campaign bed with a fixed angled headrest. The day-bed is made of solid walnut, with turned legs, and is fully upholstered. It has been made wider than would have been normal for the era, to accommodate the ample girth of Queen Anne. Of course, favoured guests and visitors would have needed somewhere to sit while in 'the presence', so four high-backed Queen Anne-style chairs with authentic horsehair padding

Day-beds

Day-beds were popular because they served as a useful upholstered seat when set along a wall, as well as offering a comfortable spot on which to rest full-length during daylight hours. Rudimentary seventeenth-century models resembled a soldier's campaign bed, essentially a long oak bench with an adjustable head-rest, topped with a bolster-like mattress. With the Restoration, and the influence of the French court, there was a market for more elaborate Baroque furniture, particularly the *chaise-longue*. New woods such as walnut were used for spiral twisted balusters, crested carvings and decorative stretchers, while seats and backs were often made from woven cane. Luxurious models such as the walnut day-bed at Belton House, Lincolnshire (right), which dates from about 1685, were upholstered and boasted a padded head-rest and foot-rest.

and appropriate upholstery were specially made for this room, to stand neatly against the newly-covered walls.

Russell turned to the Gainsborough Silk Weaving Company to supply the fabric for the upholstered day-bed, the curtains and the floor cushions. The opulent fabric chosen, a silk and cotton damask weave with a Baroque-style abstract floral pattern, woven in two shades of gold on the firm's century-old looms, was inspired by the company's extensive archives. Russell also followed the practice of former generations wishing to transform an interior by having panels of the fabric stretched and mounted over wooden battens to line the walls.

RIGHT: Eighty-three-year-old woodturner Fred Jennings turns one of the under-stretchers for the day-bed.

These fabric panels conceal the poor-quality wooden panelling in a reversible and non-destructive manner, and transform a robust and rather masculine interior into a suitable setting for a peripatetic monarch.

The finishing touch

RIGHT: The silk-lined walls of the Withdrawing Room add a feminine elegance to what was originally a rather masculine space.

BELOW: The design for the damask was inspired by patterns found in Gainsborough's extensive archives. The pattern books contain over 2000 samples.

Artist and restorer Corin Sands was commissioned by Russell to paint three pictures for Avebury Manor, all completely different in style, period and materials. In the case of the Withdrawing Room, he was asked to produce a rendition of a painting associated with Queen Anne, *Birds in a Landscape* by Jakob Bogdani, a Hungarian artist who moved to London in 1688. He was a specialist still life and bird painter at Court and the original painting is now in the Royal Collection, probably having been acquired by Anne herself. There was a fascination in Court circles with exotic birds and beasts at this time; exploration of the globe was a source of patriotic pride, and there was a burgeoning scientific interest in the wildlife to be found on other continents. Bogdani's original was painted in oils on canvas; for practical reasons Corin has interpreted this image in acrylic paint on canvas.

A Military Man
at Avebury

S ir Adam Williamson and his wife Ann – known as Nanny –
came to Avebury in 1789, when she inherited the house
from her uncle, Arthur Jones. Williamson's lengthy career
was shaped in the New World, where he served first as a
soldier and later as a British government administrator in
North America and the Caribbean. His first trip to America
with the British army was made in 1753 and he was
wounded two years later at the Battle of the Wilderness in
Virginia. Nevertheless, he returned to service and had
a number of military successes, helping to capture Fort
Henry, Louisburg and Quebec, then travelling to the
Caribbean to take Martinique and Havana.

After returning to Britain in 1763 and receiving promotion to
the rank of major, Williamson married Ann Jones in 1771. Back in
America in 1775, he fought in the first serious engagement in the
American Revolution, the Battle of Bunker Hill; his services were
rewarded with promotion in Britain a year later, and he finally gained
the rank of colonel in 1782.

Adam and Ann Williamson had only a year together at Avebury
because, in 1790, Williamson was dispatched to Jamaica to administer
and protect one of Britain's most economically
important colonies. The main crop was sugar cane,
grown on plantations owned by Britons, but
harvested by tens of thousands of slaves traded from
West Africa. On the eve of taking ship at Falmouth
in September 1790, Williamson wrote to 'Nanny'
most affectionately, reminding her that they could
send letters across the Atlantic by ship every month
until they could be reunited:

ABOVE: This portrait of Sir Adam
Williamson is by the miniaturist
George Engleheart.

RIGHT: The Battle of Bunker Hill took place on 17 June 1775
during the Siege of Boston and was an early engagement in the
American Revolution.

'This my dearest Nanny I conclude you will receive at Avebury, and I shall give you a little history of our journey here . . . restive horses, brutes of drivers & in constant expectation of going over . . . in short never My Dearest Nanny entertain a thought of going in a packett [ship], for you will be frightened to death long before you reach Falmouth . . . (though) long months seem staring us in the face, yet our constant thoughts of each other, and the collecting matter for the first Wednesday of every month will invariably lead us on to the next season, when I hope and trust I may be blessed with seeing you arrive in health and spirits.'

BELOW: The King's House in Spanish Town, Jamaica, where Williamson once resided as Governor, is now just a façade.

Life in Jamaica

The following year Williamson was appointed Governor of Jamaica and Ann sailed out to join him in Spanish Town, the capital. They lived in the governor's official residence, King's House, a Neo-classical building completed in 1762, the façade of which still stands, though the interior is in ruins as a result of a fire in 1925. The colonists and their governor lived in style, with plantation owners using their wealth and captive workforce to build large houses on their estates, but there was an ever-present fear that the slaves might one day rise up against their masters.

In September 1791 there was a violent uprising in neighbouring Saint-Domingue (now Haiti), a French colony close to Jamaica, with 20,000 killed, and British plantation owners feared the spread of insurrection. However, rumours of a slave revolt in Jamaica planned for Christmas 1791 proved unfounded, possibly because Williamson turned out the militia and the presence of 8,000 troops, including 1,000 cavalry, acted as a deterrent.

BELOW: This contemporary map, dating from 1789, shows the island of Saint-Dominge (Haiti), just two years before it erupted into violence following a slave revolt.

Williamson had a reputation for conviviality and heavy drinking, and his popularity among the 'plantocracy' of Jamaica was assured. Having gained the confidence of both the plantation owners and his masters back in London, he was rewarded by the British government with a knighthood. However, his subsequent appointment as Governor of Saint-Domingue ruined his career. Following the outbreak of war with France, the British had seized the opportunity to take control of some of the colony but relationships among the plantation owners, some of whom had backed the French Revolution, were fractious; meanwhile, they had a huge and malcontent population of slaves threatening violence. Unlike relatively stable Jamaica, Saint-Domingue was a difficult place to govern and Williamson presided over military losses, guerrilla action, massive costs to the British government and innumerable deaths of soldiers through disease.

Death and disillusionment

In addition, Williamson's beloved wife Ann died of yellow fever in 1794, aged only 48. She was buried in St Catherine's Cathedral, Spanish Town, where her ornate carved monument records that 'She was an ornament of Society and a pattern to her sex . . . her life a blessing, her death an irreparable loss to the community.'

Perceived as a 'lax administrator and poor judge of men', and presumably disillusioned with his lot, Sir Adam retired from the colonies in 1796 and was rewarded with promotion to the rank of lieutenant-general in 1797. He returned alone to England. Despite a military career spanning 45 years spent in various hazardous locations, Sir Adam expired in 1798, several days after a fall from a chair in the dining room of Avebury Manor.

ABOVE: An ornate monument to Anne Williamson, who died in 1794.

THE GOVERNOR OF JAMAICA'S DINING ROOM

This room was originally built around 1600 by Sir James Mervyn and was remodelled nearly a century and a half later, possibly for the second Sir Richard Holford (although the work might have been carried out for a tenant), with Palladian details added to the existing structure, including carved and pedimented doorways, a bold entablature and a dado. As a result, it has an overtly symmetrical arrangement typical of the eighteenth century.

During the last years of that century, the manor was under the ownership of Sir Adam Williamson and his wife, Nanny, who were mostly absent in the West Indies where Sir Adam was Governor of Jamaica. After he came back a widower in 1798, he endeavoured for a while to

BELOW: This photograph of the Dining Room during the 1920s shows the walls covered in rich flock wallpaper.

EXERCISE CHAIRS

Dining rooms of this period sometimes contained an exercise chair, in which gentlemen could keep fit without undertaking the rigours of riding a horse in inclement weather – hence the alternative name of 'chamber horse'. John Wesley, the founder of Methodism and an exercise enthusiast who lived to the advanced age of 89, is known to have owned one. Operated by spring action, such chairs worked by grasping the arms and alternately pushing upwards and sinking down to create a concertina effect in imitation of the movement of a trotting horse. The one seen here sits in the library at Belton House, Lincolnshire.

entertain at the house, having his nieces to stay and inviting officers stationed nearby to meet them. But that October he fell from a chair in the dining room – probably succumbing to a stroke – and died soon after.

The Avebury Inventory, compiled in 1798, shortly after Sir Adam's death, is an invaluable handwritten guide to the contents of the house. Among the items listed in this room were a pair of portraits of King George III and Queen Charlotte after the style of Sir Joshua Reynolds, a parlour organ, a versatile dining table in several parts with spare leaves to accommodate a large group, and a variety of seating from dining chairs to a sofa. This was evidently a room in which Sir Adam entertained company. A close reading of the Inventory, however, hints at the time he had spent abroad; the festoon curtains listed in the document, for example, though originally highly fashionable, would have been notably out of date by 1798.

ABOVE: The Inventory, dating from 1798, lists items in the Dining Parlour, including a Wilton carpet and various pictures.

GEORGIAN DINING

By the 1790s, any gentleman of substance had a dining room in which to entertain his neighbours. To be genteel necessitated a degree of hospitality and dinner was an opportunity to show off evidence of one's wealth with the furniture, the silver, the paintings and the new porcelain. In addition, by commanding a menu of gastronomic delights, the host could display his sophistication (or that of his cook).

The fashionable time to eat had settled at 7pm. Formal dinners usually comprised two or three courses, each consisting of a large number of different dishes placed on the table simultaneously, a system known as *à la française*. The diners helped themselves to the nearest dishes, while the servants replenished wine glasses. Afterwards, dessert would be served, usually a *tour de force* of sweetmeats and ices, jellies and syllabubs. The ladies retired to drink tea in the drawing room, leaving the men to pass the port and compete in telling sporting tales and fruity anecdotes. Even Rochefoucauld, no prude, remarked: 'Very often I have heard things mentioned in good society which would be in the grossest taste in France.'

ABOVE: The Dining Room at Attingham Park, Shropshire, is seen here laid out for a formal Georgian dinner, complete with candelabra and table decorations.

Company shocked at a Lady getting up to Ring the Bell.

LEFT: A cartoon by James Gillray, entitled *Lady Getting up to Ring the Bell*, pokes fun at the appalling table manners displayed by a group of suitors as they scramble to stop the lady carrying out her purpose.

TRANSFORMING THE DINING ROOM

'If I had a big house like this, I'd want all my friends to come and enjoy it,' said designer Russell Sage at the outset of the Avebury Manor project. This sentiment struck a chord with the rest of the team; it was evident that the previous inhabitants of Avebury Manor wanted to show off the house, and their status as owners, to their visitors. Consequently the Dining Room has been presented as a celebration of the achievements of Sir Adam Williamson, and is set around 1798.

Sir Adam was a military man who had seen action abroad, but he was also a clubbable individual who enjoyed socialising. The apex of his career was his Governorship of Jamaica at what was a critical time in the island's history. Russell and the experts were keen that the room should reflect this aspect of Sir Adam's eventful life. They also wanted to create a 'lived in' feel, as if a meal were in progress and visitors were being drawn into the midst of Williamson's social life.

BELOW: Prior to the transformation of the Dining Room, a temporary event called Moving Out showed the room as it might have looked between owners, complete with a packing case and chairs stacked for removal.

ABOVE: The original eighteenth-century narrative wallpaper in the boardroom at the London headquarters of Coutts bank provided inspiration for the Dining Room wallpaper at Avebury.

The Chinese Wallpaper

A hand-painted Chinese wallpaper sweeps in a continuous panorama around the walls. The design is similar to an 'industry' paper, showing trade in tea, ceramics and sugar, and it narrates aspects of Sir Adam's life, having been made specially to tell his story.

There were historical precedents of influential men of affairs being given 'narrative' wallpapers custom-made in China. George Earl Macartney returned to London in 1794 after two years as Britain's first ambassador to Beijing. A friend of the banker Thomas Coutts, he gave him the wallpaper that now adorns the bank's boardroom at their London headquarters. It features many contemporary activities: the tea industry, silk production, agriculture, pottery and even Chinese opera.

Similarly, James Drummond of the Honourable East India Company spent many years in Asia and commissioned a wallpaper for himself. The Drummond paper, now in the Peabody Essex Museum in Salem, Massachusetts, depicts the *hongs* of Canton – the warehouses through which traded goods passed. It also shows the sampans and junks in the harbour, ferrying exports to the merchant ships moored at Macau. This is one of the few surviving Chinese papers to depict Westerners, perhaps even Drummond himself.

The wallpaper for the Dining Room needed a similar narrative, as though it had been commissioned as a present for the Governor.

LEFT: Presenter Paul Martin tries his hand at painting some flowers during a visit to Fromental's studios in Wuxi, China, which lies about 125km (77 miles) north-west of Shanghai.

Jamaica was a British naval port with strategic significance, but it also functioned as a port of call for merchant ships. Each of the many trading companies operating in the Caribbean would have curried favour with Governor Williamson.

The design process

Using the two historical examples, Fromental, London-based makers of exquisite wallpapers, created a panoramic design; every detail was sketched out in pencil, then hand-painted. The design exactly fits the contours and features of the room, and depicts Western trading ships leaving Chinese harbours and heading across the seas. Avebury Manor and the standing stones can be seen, while the West Indies are represented by islands dotted with palm trees.

The East India Company imported Chinese wallpapers to Britain as a sideline to its main commodities, tea, spices and ceramics. Less well-known was the West India Company, whose main commodity was sugar. It was much in demand in British beverages, especially tea, so the fortunes of these two trading companies were intimately linked, and they have both been represented in this wallpaper. There are even thumbnail portraits of the 'Western merchants' involved in this commission; Dan Cruickshank and Russell are standing on the walls of a Chinese harbour, while presenter Paul Martin, wearing eighteenth-century hunting clothes, is pictured on horseback in the Avebury section.

The colour palette is based on the papers at Coutts and the Peabody Essex Museum, though Fromental increased the depth of some of the shades, which had yellowed in the 200 years since their creation. The main ground is coloured in soft buff, and the trees and foliage are vivid greens.

ABOVE: Artist Chen Jing paints some of the figures for the Dining Room wallpaper in Fromental's studios in Wuxi, working with two brushes at the same time.

Painting and hanging the paper

Manufactured in Wuxi in China, the paper consists of 22 panels, each 95cm (37in) wide. Individual panels were made up of smaller hand-made sheets of paper, traditionally made from mulberry bark and bamboo, then treated with an application of fish glue to control the take-up of paint. The background colour was painted first, then the landscape, water and mountains in loose shading, leaving spaces for the figures, buildings and trees. The paints used are similar to gouache, a type of opaque watercolour.

The paper was delivered from China after many months of work, and was hung in the manor using a traditional starch-based glue. As each panel is unique, and the end result of so much labour, the paper-hanging process was a nerve-racking experience, but the effect is stunning. The wallpaper is very vibrant, so the rest of the Dining Room was painted in appropriate eighteenth-century colours, using Farrow & Ball and Earthborn paints. Details on the mouldings and the triangular pediments have been gilded to add an air of opulence.

Furnishing the Room

A number of eighteenth-century pieces were brought in and placed to highlight the symmetrical arrangement of this room, for example, a pair of 'lunette' or semi-circular tables flank the fireplace, each bearing a matching candelabra. The fan-backed sideboard positioned between the two pedimented doors on the east side would have been used by footmen serving dinner (the corridor outside leading directly to the kitchen). In addition, Russell commissioned sofa manufacturers George Smith to make a 1780s Hepplewhite-style sofa, covered in yellow silk.

At the centre of the room stands the dining table. To the Georgians, this would have been a versatile item which could be expanded by virtue of extra leaves inserted over the supporting extendable frame. In the Dining Room at Avebury, the table has been laid for ten people at dinner, with the tableware, cutlery and accoutrements arranged to represent a formal eighteenth-century meal. Above the table is a magnificent chandelier, apparently fitted with candles though actually powered by electricity, as candles would be too risky to use in a house of this age.

RIGHT: A section of the magnificent hand-painted narrative wallpaper in the Governor of Jamaica's Dining Room depicts scenes from Chinese life and industry.

ABOVE: Gordon Jagger of ELG at Sainsburys fixes the upright handhold to the swept arm of the exercise chair.

BELOW: A potter at Keramis in Jingdezhen, China, works at his potter's wheel on one of the fine porcelain bowls.

The Exercise Chair

While Sir Adam was a man of action in his youth he was rather a bon viveur in later years, so he would have been likely to have owned a 'chamber horse', or exercise chair, to maintain his fitness. Thomas Sainsbury of specialist cabinet-makers ELG at Sainsburys believes that examples from this era did not often survive because they were not intrinsically attractive and rough treatment made them prone to breakages. Gordon Jagger, one of ELG's principal and most experienced cabinetmakers, and Trevor Crutcher, who oversaw the making of all the firm's furniture for Avebury, tackled this unusual project, following the same proportions as a model that once belonged to John Wesley, the founder of Methodism. The seat is covered in leather and the interior is sprung to offer extra resistance to the user. The carcase, back, step, and armrests are made from reclaimed Georgian mahogany, salvaged from furniture, while the vertical poles are topped with carved finials in the shape of acorns.

The Armorial Porcelain

Of particular interest in this room, and in keeping with the period, is the specially commissioned porcelain. The British aristocracy's passion for Chinese armorial porcelain reached its height between 1720 and about 1800. Merchants would order porcelain adorned with their clients' coats of arms and other personal designs in Canton and the whole commission, from instruction to delivery, might take as much as two years. A new peerage, a marriage between two important families, or simply the desire to impress were all good reasons to order a crested dinner service, an armorial tea set, or a punchbowl illustrated with the family's estates.

Chinese porcelain was first developed over 1000 years ago and was greatly valued for its hardness, translucency and delicacy of colour. The city of Jingdezhen, considered the birthplace of this type of porcelain, has one million inhabitants, of whom 300,000 are involved in porcelain production, many of them using traditional techniques and skills.

Benjamin Creutzfeldt, a former specialist at Christie's Chinese Department who now works with local artists to create decorative porcelain through his company Keramis S.A., commissioned one of the workshops to produce a decorative porcelain tea set and a commemorative punchbowl, exactly as they would have been in Sir Adam's lifetime. Each piece was thrown and turned, then glazed and fired all white at 1300°C. They werxe then painted with enamel colours before refiring at lower temperatures. The teacups and saucers feature the blue dragon from the Williamson crest, while the punchbowl depicts the King's House in Jamaica on one side and a full coat of arms on the other.

BELOW: This teapot and cup and saucer form part of the hand-made porcelain tea service made by Keramis. It features the blue dragon from Sir Adam Williamson's crest.

ABOVE: Benjamin Creutzfeldt and Paul Martin sift through broken fragments at the shard market in Jingdezhen in the hope of finding a genuine antique piece.

A Very British Ritual

The tea set made especially for Avebury Manor reflects one of Britain's more popular rituals – taking tea. The leaves of the tea plant *Camellia sinensis* had been used to produce a beverage in China for thousands of years before they were introduced to Europe by Dutch and Portuguese merchants. Samuel Pepys recorded his first encounter with this novelty in September 1660: 'and afterwards I did send for a cup of Tee (a China drink) of which I never had drank before'; Pepys would have been more familiar with coffee.

Catherine of Braganza, who married Charles II in 1662, was the first to introduce tea-drinking at Court, instantly making it a desirable thing to do; society folk bought packets of loose-leaf tea from coffee houses so they could try this new beverage at home. Accoutrements such as delicate porcelain cups, tea-pots and silver kettles were avidly sought, and the East India Company supplied the tea-sets as well as the tea-leaves. Both green and black tea were drunk with sugar, the demand for which was one of the principal motivations for the slave trade in the West Indies.

By the eighteenth century, tea-drinking was firmly established. Dr Samuel Johnson, in *The Literary Magazine* (1757) described himself as 'A hardened and shameless tea-drinker . . . whose kettle scarcely has time to cool, who with Tea amuses the evening, with Tea solaces the midnights, and with Tea welcomes the morning.'

ABOVE: These delicate porcelain pieces in the Chinese style date from *c.*1770 and belong to a Worcester tea service at Saltram in Devon.

In 1785 the tax on tea was reduced and home consumption rose; it was now drunk at all times of day, by all classes, and increasingly with milk. The thrifty could squeeze many cups out of a small amount of leaves by continuing to add hot water to the dregs. Country houses such as Avebury Manor would order several chests of tea from London dealers annually, along with hard, rock-like cones of sugar known as 'loaves', broken off with 'nippers' to provide fine grey crystals.

By the middle of the nineteenth century, the yawning culinary gap that had developed between luncheon at one and dinner at eight in polite circles was filled by afternoon tea. It became a thoroughly British institution with elaborate social rituals that involved the hostess presiding over the teapot and chivvying male guests to hand around the tea-cups. Tearooms and teashops across the country became places where the upper and middle classes could meet in respectable public surroundings on afternoons out, while the working classes settled for 'high tea', a substantial early evening meal accompanied by the national drink.

RIGHT: A table is set for the ritual of taking tea in the Velvet Drawing Room at Saltram, Devon.

BELOW: A painting by Enoch Seeman at Belton House, Lincolnshire, shows the Cust family taking tea, a ritual that had become well established by the eighteenth century.

The Soft Furnishings

A grand setting required grand fabrics; to complement the colour scheme of the Chinese wallpaper, Gainsborough Silk Weaving Company supplied 48m (52½yd) of silk and cotton damask in a diaper pattern in gold for the upholstered furniture. The curtains were based on those designed for the Prince of Wales, later George IV, for the Chinese Drawing Room at Carlton House, London. They were designed by Thomas Sheraton, primarily known as a furniture-maker, and the design appeared in his own publication, *The Cabinet-Maker's and Upholsterer's Drawing Book*, published in parts between 1791 and 1794. The two sets of curtains for this room, topped by ornate pelmets, took up 48m (52½yd) of Gainsborough's classic green silk and cotton stripe, and all tassels, fringing and braids were provided by the specialist firm of Henry Newbery after consulting their extensive historic archives.

BELOW: A detail of the carpet's design shows the influence of Neo-classicism.

Carpeting the dining room

The Avebury Inventory lists a 'large Wilton carpet' in the Dining Room. Such items were very expensive, even by aristocratic standards, and highly prized. In the dining room at Saltram, there is a magnificent Axminster carpet made by Thomas Whitty in 1770. It cost £126, the approximate equivalent of £15,000 today. The pattern on it echoes the ornate design of Robert Adam's ceiling above.

So valuable were carpets in this era that in preparation for formal meals servants often placed a oil cloth, painted to match exactly, on top of the carpet before the dining tables were erected in order to protect against accidental spillage. To create a modern version in the Neo-classical style, Ulster Carpet Mills referred to their archive and other historical sources. Their intricate design, with its central oval motif, makes references to architectural details in the Dining Room, such as the elegant frieze and the plasterwork band of oak leaves over the doors.

LEFT: A weaver works at the loom at Ulster Carpet Mills.

BELOW: Presenter Penelope Keith and designer Russell Sage leaf through the archives at Gainsborough Silk Weaving Company.

The Finishing Touches

To complete the room, artist and restorer Corin Sands was commissioned to paint a picture of the Governor's Mansion, known as the King's House, in Spanish Town – the capital of Jamaica at the time. The grand Neo-classical façade survives to this day, an eloquent reminder of colonial rule. Using photographs and contemporary images, Corin created a suitably archaic depiction of the Mansion in acrylic on canvas.

Also on the walls, Sir Adam and his wife Nanny are represented in a charming pair of *eglomisé* silhouettes, a technique which involved painting on the reverse of a pane of glass. The silhouettes, once owned by the Williamsons, were traced to America by the researchers, and purchased for the Dining Room, returning personal images of Sir Adam and Nanny to Avebury Manor more than two centuries after the couple lived here.

OVERLEAF: The magnificent wallpaper in the Governor of Jamaica's Dining Room depicts a narrative tale that includes scenes from Adam Williamson's life, Chinese industry and British trade.

CHAPTER IV

AVEBURY IN THE VICTORIAN AGE

THE NINETEENTH CENTURY'S DYNAMISM AFFECTED EVEN RURAL WILTSHIRE, AND THE DEVELOPMENT OF SWINDON AS THE HUB OF BRUNEL'S GREAT WESTERN RAILWAY BROUGHT STEAM-POWERED TRAVEL AND CONTACT WITH THE WIDER, FASTER-MOVING WORLD. BUT FOR THE KEMMS, A FAMILY OF FARMERS LIVING AT AVEBURY MANOR, THE PREOCCUPATIONS REMAINED THE AGE-OLD ONES OF SOWING AND HARVEST. DOMESTIC LIFE WAS MODEST BUT COMFORTABLE, SUPPORTED BY A PHALANX OF SERVANTS BASED IN THE KITCHEN, THE HEART OF THE HOUSE.

LIFE IN AN AGE OF ENTERPRISE

Nineteenth-century Britain saw a period of massive expansion as increasing industrialisation turned once quiet towns such as Birmingham, Manchester, Leeds and Sheffield into major industrial cities. Newly constructed canals and railways transported mass-manufactured goods and people from one end of the country to another, and overseas trade with Britain's colonies and dependencies and other nations grew exponentially. Fortunes were made in manufacturing, banking and trade, and money was ploughed into new buildings to reflect civic pride and personal prestige. Yet alongside this

LEFT: This painting by William Bell Scott, entitled *Iron and Coal*, reflects the enormous pride taken by the Victorians in their industrial achievements.

VICTORIAN TIMELINE

1837
Publication begins of Charles Dickens' *Oliver Twist*

1842
Swindon Junction Station opens on the newly built Great Western Railway, vastly improving journey times between Avebury and London

1837
Victoria, George IV's 18-year-old niece becomes queen; within a few years people are referring to the era as 'Victorian'

1840
Work starts on the new Houses of Parliament, rebuilt in Gothic style following a devastating fire

1851
Britain stages the Great Exhibition – the first world's fair

wealth creation came poverty of a kind never witnessed before, as poorly paid factory workers were crowded into squalid, badly built housing, devoid of sanitation, privacy and all hope.

A new era begins

Meanwhile, the royal succession was assured in 1837 by the accession to the throne of a diminutive but determined eighteen-year old, Victoria, who set her sights on marrying her cousin, Prince Albert of Saxe-Coburg-Gotha. Their marriage provided a model of domesticity, fecundity and fidelity that coloured the moral tone of the nation for the rest of the century. Britain's dominant role as the world's 'most advanced nation', reigned over by a respected royal family, was highlighted by the phenomenal success of the Great Exhibition of 1851, a commercial triumph headed by the Prince Consort. This event was a showcase for the world's modern nations to display their manufacturing skills and talent, and was housed in an innovative new building which *Punch* magazine nicknamed the Crystal Palace.

ABOVE: Queen Victoria and her Consort, Prince Albert, were keen to be portrayed with their ever-expanding family, as a way of promoting family values.

1854
The Crimean War begins, the only European war fought by Britain during Victoria's long reign

1868
This same year, Benjamin Disraeli and William Gladstone become Prime Minister for the first time

Disraeli

1880
Cragside, in Northumberland, becomes the first house to be lit by electricity using Joseph Swan's incandescent light bulbs

1861
William Morris establishes a decorative arts firm which ultimately becomes known as Morris and Co.

1875
Liberty opens on Regent Street, London, selling exotic goods from Japan and the East

Culturally, there was a fervent interest in scientific discovery and reasoned research. Darwin's theory of evolution, propounded in *On the Origin of Species*, was so radical that when it was published in 1859 it caused a storm of debate thanks to its perceived irreligiousness. Technological innovation also made a radical impact on people's lives. Developments unimaginable in 1800, such as gas lighting, mass transportation, anaesthetics and photography were commonplace by mid century, only to be trumped by the end of the Victorian era by the motor car, moving pictures and electricity. It was an era of complete social and political change.

ABOVE: Completed in 1894, Standen in West Sussex had one of Britain's earliest electrical systems. This original hall light was designed by the Arts and Crafts metalworker W.A.S. Benson.

BELOW: *The Bayswater Omnibus*, by George W. Joy, shows a group of passengers from different social classes taking advantage of one of London's most popular modes of transport.

THE GREAT EXHIBITION

The apogee of Victorian pride was the Great Exhibition of 1851, a celebration of the 'Works of Industry of all Nations'. It was housed in an innovative iron and glass building in Hyde Park, London, designed by Joseph Paxton, the head gardener at Chatsworth, and was the first international exhibition of manufactured products. The event proved hugely popular, with over six million people passing through the turnstiles. But not everyone was impressed: art critic John Ruskin disparagingly called the building a 'great cucumber frame' and a young William Morris was so appalled by the whole idea that he refused to go.

Nonetheless, the exhibition was enormously influential. After it had closed, one of the organisers, Henry Cole, collected the best items in the show and put them on display in what was to become the Victoria and Albert Museum, to inform and educate future generations of designers and craftspeople.

The Avebury backwater

Throughout the nineteenth century, however, Avebury remained something of a backwater. The landowners and their tenants weathered successive agrarian depressions, and celebrated good harvests when they had them. The manor was occupied throughout the late Georgian and Victorian eras by a reasonably prosperous but hard-working family of tenant farmers, the Kemms, who took up residence in about 1816, renting the house and farming the land until 1902, when they relinquished the lease. Throughout this time, the manor continued as the hub of a busy working agricultural estate, as well as a family home, and the fabric of the house became slightly dilapidated during decades of benign neglect.

THE VICTORIANS
AT HOME

During the Victorian era, cosy domesticity, as personified by Victoria and Albert, became the ideal for British families, particularly the middle classes. A Victorian gentleman was expected to shoulder the social responsibility of creating a home and, ideally, a full nursery. His purpose-built villa was a complex structure, with an evident hierarchy of rooms reflecting the status of its various inhabitants. It was in the nineteenth century, too, that rooms in grand houses gained new and specific purposes, for example, the gunroom, the billiard room, the study and the smoking room.

While rooms such as these were designed for strictly masculine enjoyment, there were cosier, more domestic rooms where the mistress of the house, with servants to relieve her of any domestic toil, pursued craft activities such as shellwork, crochet, or decoupage (the art of sticking colourful cut-out illustrations onto household objects for decoration, often with surreal effect). Charity work provided excuses for gentlewomen to escape the house, which was otherwise regarded as their 'proper place', as did 'at homes' – a ritual of calling on one another for tea and company.

ABOVE: Making decoupage screens became a popular pastime in the Victorian era. Paper cut-outs depicting flowers, children, animals and country scenes could be bought in packets and stuck onto ready-made screens. This one is from the hall at Bradley Manor, in Devon.

Home comforts

Throughout most of the nineteenth century a greater complexity of furnishings was the fashion. Wallcoverings were transformed in 1841 with the development of machines to print wallpaper. Power looms generated fabrics more swiftly and cheaply than ever before, and the Victorian sitting room became a riot of fabrics and patterns, tassels and pompoms, stifling luxury and dust-gathering opulence. Over-stuffed upholstery was topped with lacy antimacassars; riotous surface patterns vied with novelty objects, such as stuffed animals or wax flowers. Soft chairs were sprung to make them more comfortable, and an absence of arms on chairs allowed the sitter's crinoline to be accommodated.

The Victorian sitting room became a riot of fabrics and patterns, tassels and pompoms, stifling luxury and dust-gathering opulence.

New materials were used for furniture – brass, for example, became popular for bedsteads because it was considered more hygienic than wood – while innovative techniques created fresh possibilities, such as bentwood chairs. The Victorians took a particular delight in using one material to imitate another, such as moulded papier-mâché furniture that appeared to be lacquered wood, and ceramic vases that looked like bamboo.

The development of aniline dyes in the 1860s and 1870s changed the prevailing colour palette; now it was possible to produce strong, dark purples and vibrant fuchsias, deep, resonant reds and virulent greens. Both in dress and furnishing fabrics, there was an explosion of vibrant colour previously unattainable. However, darker colours were also favoured as they were less likely to show the dirt.

BELOW: The smoking room at Lanhydrock, Cornwall, displays the Victorians' love of patterned surfaces, rich colours, comfortable well-sprung seating and ornaments.

ABOVE: Wightwick Manor in the West Midlands was built and furnished under the influence of the Arts and Crafts Movement. The galleried parlour is designed to look like the Great Hall of a medieval manor house.

Exoticism versus tradition

The prevailing passion for novelty and ingenuity was reflected in the styles of contemporary interior design. The Victorians were curious about other cultures, such as the classical world, ancient Egypt, India, China and Japan (formerly a 'closed country' that only began to trade with the West in the middle of the century). Imports of exotic goods were available through new department stores such as Liberty, but contemporary British designers and manufacturers also responded to the demand by imitating foreign models. The plethora of styles on offer to the householder was staggering; books such as Owen Jones's *Grammar of Ornament* (1856) offered useful guides to the available styles. Many households indulged in exuberant eclecticism and Victorian parlours were tributes to the concept that 'more is more'.

Looking to the past

Alongside the interest in the new and the exotic, a fascination with historical revivalist styles also swept through Britain. Partly as a reaction to the mass industrialisation of the age, the newly wealthy middle classes were keen to show their cultural aspirations by favouring styles of former eras. In effect, they were buying history. The nostalgic hankering for the manners and morality of the Middle Ages led first to the Gothic Revival and finally found full expression in the Arts and Crafts Movement, embodied in the passionate and dedicated design reformer and poet, William Morris, who had dismissed the Great Exhibition as 'tons upon tons of unutterable junk'.

Yet in rural backwaters such as Avebury these changing fashions caused barely a ripple. Traditional crafts and styles still survived, and while the Kemm family, as prosperous farmers, would have bought some of the mass-produced trappings of respectability for their parlour, the majority of utilitarian domestic furniture was still made locally by craftsmen in vernacular styles and materials.

WILLIAM MORRIS

Having by his own account read the works of Sir Walter Scott by the age of seven, William Morris (1834–96) grew up with a passion for the past and a horror of all things modern, including mass manufacturing and the factory system. He urged people to 'Have nothing in your houses that you do not know to be useful or believe to be beautiful' and, through his company Morris and Co., set out to reform the manufacture of household furniture and furnishings by reviving traditional craft skills and creating designs inspired by nature and the arts of the Middle Ages.

In 1871 Morris took out a lease on Kelmscott Manor, on the River Thames near Lechlade in Gloucestershire, a seventeenth-century gabled stone house that bears a remarkable resemblance to Avebury Manor. The house became Morris's architectural ideal, and it was here he spent his summers away from the hustle and bustle of London life, seeking inspiration for his designs from the willow-lined walks along the river and the surrounding countryside. It was also the model for the 'old house by the Thames' in Morris's Utopian novel, *News From Nowhere* (1890), which envisages England transformed into a communist paradise, where people are free and equal, and industrial squalor and central government are things of the past.

ABOVE: The frontispiece of the 1892 Kelmscott Press edition of *News from Nowhere* depicts Kelmscott Manor – the focal point of Morris's story.

BELOW: William Morris is possibly best remembered today for his wallpapers. 'Daisy' was one of his earliest and most popular patterns, widely used for the bedrooms of maids and young girls.

SERVANT LIFE

The Victorian household was reliant on dependable servants, working long hours every day in return for their bed and board and a small salary. In rural areas such as Avebury, there was no shortage of young people wanting to work at 'the big house'. Though the work was physically demanding, many youngsters were escaping life with exhausted parents in a rundown cottage with too many siblings and an uncertain income. To a trainee kitchen maid or hallboy, living in a well-appointed house with plenty of decent food, even if one had to share a bed with another lowly servant, was a great improvement. There was security, a sense of one's place in the social strata of the household and a modest income; in time, there would be promotion, a pay rise and a change in status.

Much of the daily workload of domestic servants was repetitive drudgery. In an era of open fires in every room, candles or paraffin lamps for lighting and fairly primitive cooking facilities, the main problem was how to keep a place warm, well-lit and clean. Until the middle of the nineteenth century, water was usually pumped by hand from the well outside in the courtyard; linking the supply to a tap in the kitchen where it was most needed was a great improvement. Soap was sold in large bars by grocers and, to make a detergent solution suitable for washing dishes or cleaning floors, the bar would be grated by hand – a tedious process.

A working day

A great deal of a servant's working day was spent on cleaning, from the kitchen maid black-leading the range every morning to the parlour maid raking out the ashes from fireplaces and making up new fires. The kitchen table would be scoured with sand and hot water, while the scullery maid would clean and peel the muddy vegetables. While

ABOVE: Thanks to the widespread use of bellboards, servants were constantly at the beck and call of their employers. This one sits in the servants' passage in Erddig, near Wrexham in Wales.

BELOW: A fire blazes in the kitchen range at Townend in Cumbria. The early part of a maid's morning was taken up with blacking the range, cleaning the grates and making up the fires throughout the house.

the family was breakfasting, a maid would take the opportunity to remove used chamberpots from the bedrooms, dispose of the contents in the outside privy and replace the clean pots for next time.

Until the middle of the nineteenth century, when the technology to introduce piped water throughout the house became available, servants in all households carried jugs of hot water upstairs to fill the tin baths in the bedrooms and then removed the dirty water by the same method afterwards – laborious work. In rambling and idiosyncratic old houses such as Avebury Manor, the prospect of introducing new-fangled technology to improve the quality of life of the residents was a daunting and expensive prospect. While there were still ample numbers of servants to 'do the rough work', many householders saw no need for innovations.

BELOW: The dressing room at Carlyle's House in London shows how primitive bathing conditions were until the introduction of piped water and bathrooms. It was a servant's lot to carry jugs of hot water upstairs to fill the tub.

SERVANTS AT AVEBURY

Avebury Manor was a working farmhouse, so the womenfolk of the Kemm family would have been more involved in the practical running of the household than would have been the case in a grander setting. Nevertheless, each member of the household staff would have known their place in the hierarchy beneath the family, indoor servants invariably considering themselves more refined than 'outside' ones. The kitchen staff would have frequent contact with the outside staff – the home farm workers such as dairymen who brought in milk daily, or the gardeners who would supply fresh fruit and vegetables from the extensive kitchen garden on site.

The most senior of the domestic servants in a Victorian household such as Avebury Manor is likely to have been the cook, who was a figure of some authority and took her instructions directly from the mistress of the house. The cook would deputise a kitchen maid to produce the meals for all the servants. This more junior member of the kitchen staff would learn to cook on the job and in time would hope to be employed as a cook in her own right. She had normally graduated from a lowly role as a scullery maid, forever with her arms in cold and dirty water, scrubbing vegetables, or in hot and soapy water, washing up.

House staff

In addition to the kitchen staff, there were the 'house' staff. Avebury was not grand enough at this time to employ a butler or footmen; in any case, it was expensive to have male servants as they required livery and there was a government tax imposed on their employer. Instead, two parlour maids would have been sufficient to run the house, deal with visitors, keep the fires lit and the rooms clean, and wait at table. Occasionally parlour maids might have to act as ladies' maids as well, helping the womenfolk of the family to dress and do their hair. Because they spent more time in close proximity to the gentry, parlour maids considered themselves more sophisticated than most other members of the indoor staff.

ABOVE: In *Her First Place,* artist George Dunlop Leslie found a parlour maid's plight a suitable subject for a sentimental picture.

RIGHT: The kitchen at Wallington, Northumberland, is arranged just as the servants might have left it in 1900.

ABOVE: A cook might have a huge range of pots, pans and other cooking utensils, known as a *batterie de cuisine*, part of which is seen here in the kitchen at Attingham Park in Shropshire.

The Victorian kitchen

One area that did benefit from technological advances in most households was the kitchen. The development of reliable, cast-iron cooking ranges in the 1840s was a great improvement over the previous arrangements – usually a combination of roasting meat on spits over a fire and cooking in pots on top of the hearth. Cooking ranges ran on solid fuel and did not have thermostatic controls, but they could now provide large quantities of hot water through the 'back boiler'.

There was now a wider variety of equipment to help the kitchen function more effectively, from copper saucepans in every possible size and shape to whisks and graters, mincers and choppers, ice-boxes and decorative jelly moulds. Sturdy white ceramic basins and plates were usually decorated with a K to ensure they did not stray from the kitchen. Essential kitchen equipment was displayed on open shelves or wooden dressers, immediately to hand.

At Avebury Manor, there is an anteroom off the kitchen which acted as a common room for the seven servants in residence by the beginning of the twentieth century. Along with the kitchen, this would have been the focus of the servants' daily activity.

While purpose-built Victorian houses tended to place the kitchen in the basement if possible, Avebury Manor has its kitchen and anteroom on the ground floor – though it is evident that the kitchen wing was originally placed at a judicious distance from the rest of the complex to reduce the risk of fire spreading throughout the buildings.

LEFT: The anteroom at Avebury leads directly off the kitchen, and would have acted as a common room for all the staff resident at Avebury at the beginning of the twentieth century.

LANHYDROCK

One of the best places to see a classic Victorian kitchen is at Lanhydrock, a Cornish country house which was largely rebuilt by the Robartes family in the 1880s following a disastrous fire. The kitchen complex was designed around a central corridor, with all produce arriving at the 'back door' to be stored or processed in specialist rooms such as the larders for meat, fish and dry goods, the butchery, the bakehouse and the water-chilled dairy before progressing to the main kitchen.

An army of staff worked hard at the wooden tables laid out in the vast kitchen to produce food for the family and their guests. All the equipment they might need was on display on open-shelved dressers and cabinets. From the adjacent service rooms came everything that might be required by the chef, from vats of decoratively prepared vegetables to dozens of ice-cream sundae dishes.

Meals were dispatched by relays of servants carrying laden trays away from the serving hatch, along a dog-leg corridor to reduce the transmission of cooking smells and through a green baize door into the dining room antechamber. Here food could be kept warm in heated cupboards until the family was ready to be served.

BELOW: The kitchen at Lanhydrock was fitted out with all the latest equipment. At the far end of the room are the spits for roasting meat, which were powered by a large fan fitted in the flue above the fire.

A Farming
Family at Avebury

During the nineteenth century Avebury Manor and the land around it was leased to the Kemms, a farming family. William Kemm was born in 1781 and married Jane Canning in 1807. According to one story, they were living on a farm in the Marlborough Forest when he bought a lottery ticket on a whim and gave it to Jane for safe-keeping. She sewed the ticket into her stays and forgot about it. Three months later, they discovered they had won a total of £2,500 (worth approximately £90,000 today). The Kemms negotiated with the agent acting for Richard Jones, who owned Avebury Manor and farm, and moved into the fully furnished house in 1816; the family lived there for more than 85 years.

On William's death in 1846 the tenancy passed to his third son, Thomas. He had considered studying law, but recorded in his diary that: 'I inclined such a taste for country life with all its natural beauty, the song of the birds, the perfume of the flowers and the splendour of the heavens night and day, that I would never have been happy barred

Below: The Kemm family gathers for a photograph outside Avebury Manor in about 1890.

from such delights.' Thomas lived at Avebury Manor for virtually all his 83 years and, although the estate changed hands in 1873, when it was bought by Sir Henry Meux, the lease was honoured.

Thomas married his cousin Everdell Matilda Canning in 1848, but she died in 1863, aged only 36. Lengthy mourning was fashionable in Victorian high society, but on working farms widowers tended to be more pragmatic. Thomas, then aged 44, married the 24-year-old Ellen Elizabeth Sainsbury as soon as possible in order to provide his five young children with a stepmother.

ABOVE: During the Kemm era, the state bedchamber at Avebury was cluttered with furniture dating from different periods.

A man of substance

In the census return for 1861, Thomas appears as a farmer of '1260 acres' (510ha), employing 20 men and 17 boys. The agricultural depression of the 1870s took its toll; the 1881 census records him as a farmer of '850 acres' (344ha) with fewer employees. Nevertheless, the Kemms were prosperous; the children were well-educated, with two sons becoming clergymen and both daughters teaching local children to read and write in the Avebury National School. Thomas served as church warden for more than 50 years and was interested in archaeology; in 1879, 150 members of the Wiltshire Archaeological Society 'took luncheon in Mr Kemm's large Barn under the presidency of Sir John Lubbock, visiting the Temple, the Church and the Manor'.

Thomas Kemm died in 1899 and was buried in the churchyard of St James at Avebury. He had been a popular and convivial figure, willing to show strangers around the Manor, according to the *Devizes Gazette*. In 1902 the Kemm family relinquished the tenancy and the contents of the house, undisturbed for a century, were sold in a two-day sale. Among the 'many fine things' to go under the hammer was, reputedly, an oak refectory table dating from the Manor's monastic past, which sold for sixteen guineas (£16.80).

BELOW: A catalogue from 1902 lists all the items on sale at Avebury Manor.

THE KITCHEN AT AVEBURY MANOR

The Avebury Manor kitchen and adjacent anteroom are very old. The kitchen itself forms part of the original Tudor house, confirmed by recent tree-ring dating of the timber beam above the large fireplace, which suggests a felling date of between 1550 and 1580. Expert Dan Cruickshank believes this room formed part of the manor's single-storey Great Hall and was once united with the parlour and lobby to the south. The existing partition wall around the parlour is more recent, possibly dating from the seventeenth century.

The kitchen forms part of the original Tudor house, confirmed by recent tree-ring dating of the beam above the large fireplace.

By modern standards the kitchen doorways are exceptionally low and wide, and there are enormous oak beams supporting the ceiling. During the Victorian era, this room would have been a hive of activity, lying at the heart of the Kemm household.

BELOW: The kitchen before its transformation, with its built-in dresser and curved china cupboard.

The large fireplace is a very early feature of the house, and from the Tudor to the Edwardian eras this is where food would have been prepared. Cooking evolved from a largely unpredictable business of spit-roasting over an open fire to the advent of cast-iron ranges with ovens, usually placed in the hearth because of the existing chimney flue. Though easier to regulate, and useful for their provision of constant hot water, these behemoths needed to be black-leaded every day to keep them from rusting, and they often consumed more than a ton of coal a month.

The well-equipped Victorian and Edwardian kitchen typically had a central worktable laden with cooking implements and ingredients and a well-stoked range giving off heat on one side. Crucial to health and efficiency was a washable floor, as it might need swabbing several times a day. Mass-produced glazed-earthenware floor tiles were a Victorian innovation, replacing the earlier flagstones to be found in places such as Wallington in Northumberland. Simple, wipe-clean wooden surfaces were similarly kept clean by the application of grated soap, scalding hot water and a bristle brush – all applied with plenty of elbow-grease, in an era before rubber gloves.

ABOVE: An old photograph from about 1900 shows the kitchen dresser as it would have looked during the Kemms' tenancy of the manor.

RECIPES AND RECEIPTS

Victorian cooks jealously guarded their knowledge. They collected recipes and tips, transcribing them into notebooks such as the 'Book of Receipts' compiled by Mrs Hale Parker, the cook at Arlington Court, Devon in 1876. She recorded household remedies for cleaning stained marble or deterring moths alongside alarming recipes such as Sheepshead Soup. The nineteenth century also saw the publication of various tomes on the art of cookery; Mrs Isabella Beeton's compendious *Book of Household Management*, published in 1861, systematically tackled every aspect of domestic management, including advice on servants' wages, the wearing of mourning, and, naturally, recipes.

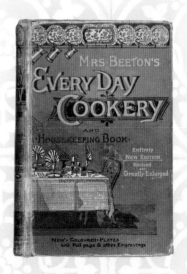

MISTRESS OF THE KITCHEN

The successful Victorian cook was a competent strategic planner, managing the provision of large quantities of cooked meals for the whole household. She required considerable physical stamina as well as patience to master the idiosyncratic cooking range. She kept kitchen skivvies up to the mark, charmed recalcitrant gardeners and negotiated discounts with local tradesmen; above all, she had to report daily to 'the mistress', her employer. The quality of her cooking, though important, was almost incidental to her organisational abilities.

The Servants' Magazine of February 1856 published 'Cook's Directions', a rigorous list of the onerous duties of a cook, stressing the value of frugality, avoiding waste, conserving fuel, and caring for pots, pans and utensils so that they did not need replacing often. Clean kitchens, punctuality and adequate supplies of flour, tea, condiments and spices were necessities. In addition, the readers were warned: 'A cook must be especially neat in her person and dress; she must avoid dirty

BELOW: The Victorian kitchen at Wallington, Northumberland, with the table set for making preserves and fruit pies.

hands and straggling hair, untidy caps and aprons, and a slatternly appearance, which no press of business will justify.'

Such admonitions, appearing in a trade journal for literate servants, indicate that the professional standards of Victorian cooks were highly variable. In fairness, most cooks were expected to work six and a half days a week, in a workplace that was excessively hot, badly lit, ill-equipped and inconveniently laid out, making the maintenance of a neat appearance a challenge at times. Many cooks grappled with antiquated equipment well into the twentieth century; some became mutinous and others took to the bottle.

Cooks' rules

Every morning the mistress and cook discussed the meals for the day, and for the forthcoming week. In many households the cook provided her 'books' – bills and accounts from the butcher, grocer and fishmonger – for the mistress's approval. Typically, cooks resisted any attempt by the mistress to influence the cost of the raw materials; they cultivated their own relationships with tradespeople, many of them mutually beneficial, and resented any interference.

Cooks were always referred to as 'Mrs', regardless of their marital status. They could be touchy, and needed careful handling if they were to be retained; they insisted on autonomy in the kitchen, even requiring advance notice if the mistress wished to visit their domain. Nevertheless, a decent cook was essential to the efficient running of the Victorian house, and was genuinely missed on her departure.

ABOVE: A good cook, like this one portrayed in an 1890 advertisement, was essential to the running of the Victorian house.

ABOVE: Victorian copper jelly moulds at Penrhyn Castle, Gwynedd. The duties of the cook included looking after pots, pans and utensils so they didn't need replacing often.

TRANSFORMING THE KITCHEN

A number of early photographs have survived, portraying the kitchen in some detail. These, combined with research from other historic houses, greatly informed the design decisions. The plan was to remove the modern Aga and basin; to restore the existing historic fixtures and make good the surroundings; to take up the 1980s flooring in the hope of revealing the original stone floor underneath (which sadly proved not to be the case); and to reinstate appropriate equipment and furniture using the photographs as a guide.

One of the items reinstated with reference to the photographic evidence was a large-scale flour bin. Considerable quantities of flour were needed in country houses as bread was baked daily, so a secure container was essential for keeping flour dry and away from vermin.

The kitchen range

Central to the running of both a Victorian and Edwardian kitchen was the range, and in transforming this room it was imperative to find one that matched the one in the photograph (shown on page 190), a coal-fired range with a number of ovens. Neville Griffiths, an expert in architectural salvage, found a very similar model, and managed to dismantle it and remove it from a house in the Wirrall shortly before the property was demolished. Like the original seen in the photograph, this was a cast-iron Wellstood range, made *c.*1904 by the prestigious Glaswegian company of Smith and Wellstood. The range was a high-quality model in its day, with a back boiler to provide hot water. Neville restored it to working order and installed it in the kitchen.

ABOVE: Neville Griffiths studies his collection of original architectural journals to help him track down the right sort of range for Avebury.

Freestanding and Fitted Furniture

Russell was keen to bring in appropriate furniture to represent the kind of practical, timeless pieces to be found in servants' quarters, and acquired a robust pine table with a drawer at each end for cooking implements and tools. Around the table are four traditional wooden chairs, simple and robust. The surviving work surfaces are much lower in height than we would find comfortable nowadays. In the nineteenth century the kitchen staff were often barely adolescent, nearly always female, and had often grown up in poor families, so were physically underdeveloped.

The shelves of the magnificent wooden dresser which runs the length of one wall were, according to a photograph from around 1900, stacked with big pewter platters and serving plates, a selection of copper saucepans and lids in ascending order of size, and a china dinner service, complete with lidded tureens, plates and a tea service. This venerable piece of furniture received a simple but thorough clean and polish, and the lower cupboard doors were replaced as they were not originals. Then the dresser was laden with appropriate kitchenware, including copper pots and pans bought by Russell. Local people also contributed to fitting out the kitchen; following an appeal for arcane kitchenware, Avebury Manor was supplied with butterpats and bottles, an authentic Victorian mincer and assorted crockery and cutlery. There were also gifts of kitchen utensils such as cast iron kettles, and a battery of labour-saving

ABOVE: Neville wanted to maintain as much of the time-worn patination on the hob plate as possible, so, using a cotton polishing mop on a buffing tool he gently brought the surface back to life.

BELOW: The resurrection of the range involved assembling all the pieces and putting them back together in the correct order. Here, the cast iron flues are being attached to the back of the range.

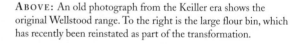

ABOVE: An old photograph from the Keiller era shows the original Wellstood range. To the right is the large flour bin, which has recently been reinstated as part of the transformation.

BELOW: A Pots and Pans Day organised by the BBC resulted in several early pieces of kitchen equipment being donated to the Avebury project.

devices, including cheese graters and a number of decorative jelly moulds.

There is a barrel-backed china cupboard in the kitchen, which was probably designed for a parlour or drawing room. It probably dates from the eighteenth century and is very similar to one at Lytes Cary, Somerset where Leopold's brother Walter and Nora's sister Flora lived. The cupboards may well have been a pair collected on one of the Jenners' many forages for architectural salvage. The location of this cupboard, at a right angle to one of the walls, initially seems puzzling, but it may have been installed in order to deflect any draughts from the corridor reaching the staff sitting next to the fire.

Floor and Walls

At some point modern quarry tiles had been laid on the kitchen floor. The National Trust sought listed building consent to have them taken up, but unfortunately the process did not reveal original flagstones as had been hoped – just a thick layer of cement and a crucial waste pipe. This ruled out the possibility of installing heavy stones, so it was decided to lay down linoleum, a material patented in the early 1860s and well-suited to hard-wearing floors that needed frequent washing. Originally the walls would have been whitewashed to keep them looking clean, and mouse traps would have been set in corners to catch questing rodents. Contrary to city practices, country houses often deterred the staff from keeping a cat to tackle rodents; they had a tendency to capture and bring in wild mice to play with, which exacerbated the problem if the prey escaped.

LEFT: The fully restored kitchen range, which is similar to the model seen in the photograph opposite, looks splendid in its new setting.

OVERLEAF: Once more the oak dresser is filled with kitchen paraphernalia, as if one of the Kemm's servants had just this minute finished tidying it up.

THE KITCHEN GARDEN

David Howard, formerly Head Gardener at Highgrove for more than a decade to HRH the Prince of Wales and HRH the Duchess of Cornwall, says 'horticulture is the biggest subject in the world.' He leads a double life; he and his family live on Lees Hall Farm in Northumberland, where he raises Belted Galloway cattle and free-range hens. He is also a very well-respected horticultural consultant, travelling the country and overseas to dispense advice on gardens to his numerous clients.

The challenge of restoring the kitchen garden at Avebury Manor appealed to David from the outset. There had been a considerable kitchen garden on this site in the early twentieth century, but by 2011 it was a wilderness of waist-high vegetation and opportunistic bushes, concealed behind two rather unprepossessing wooden doors. The combined muscle of 40 dedicated local volunteers of all ages cleared the detritus, revealing a roughly rectangular pitch, with three very high brick walls and a hedge formed of a screen of trees at the west end.

Designing the kitchen garden

Drawing on his considerable experience, David has designed the kitchen garden from scratch, taking as his inspiration the type of kitchen garden typical of country estates in the 1880s and 1890s. As much as possible, the garden includes elements and features typical of the era, from a traditional potting shed to a Victorian glasshouse. The intention is to grow a mixture of ornamental and unusual vegetables, fruit trees including nectarines, pears and apples, and flowers for decorating the house. The glasshouse will provide protection for delicate plants and there are cold frames for hardening off young plants.

ABOVE: David Howard and some volunteers work on the newly created raised beds.

LEFT: The somewhat overgrown kitchen garden at Avebury Manor before its transformation.

ABOVE: An espaliered fruit tree has been trained up the north-facing wall.

BELOW: The army of volunteers setting off to work in the kitchen garden.

Traditionally, nineteenth-century kitchen gardens were made up of a complex of raised beds, accessible to the gardeners and linked by wide walkways. In keeping with this, David came up with two modules – based on traditional, imperial measurements – which seemed ideal for the site. He has created a network of walkways, using traditional 6ft- (1.8m-) wide paths that allow a person to turn a laden wheelbarrow a full 360 degrees without difficulty. These paths link a rectilinear arrangement of beds measuring 16ft sq (4.8m sq), with narrower beds along the boundary walls. The 16ft (4.8m) measurement was a unit commonly found in Britain's agricultural history, being the turning circle required when ploughing with a team of oxen, and is the basis from which our traditional acre is derived.

While creating the paths, the workforce dug and retained the valuable topsoil, adding it to the enclosed beds. The paths were then topped with hoggin – a hardwearing conglomerate of rolled rough sand, clay and gravel typical of the Victorian era – and cambered to avoid muddy puddles forming.

A single pine has been left at the western edge of the garden, partly as an ornamental touch. 'Every garden should have one eccentricity,' says David. The plan is to raise the crown of the pine through pruning and take advantage of the shelter it will provide for shade-loving plants, among which are a wide variety of British native plants and vegetables.

To add to the visual appeal, the gardeners have added espaliered fruit trees including several morello cherries, to the curved north-facing wall. In the centre of the garden is a large square of lawn, with a weeping purple

VICTORIAN HERBS

British households had long grown herbs and spices in a spot easily accessible from the back door because they were required virtually every day. Most country houses had a stillroom – a place where fruit was bottled, pickles made, and cordials, liqueurs, medicines and perfumes distilled by the mistress or housekeeper.

Rudimentary painkillers such as headache and fever remedies were extracted from meadowsweet or willow bark, both of which contain salicylic acid, the principal ingredient of aspirin. Camomile tea calmed the nerves; a peppermint infusion soothed the stomach and sweetened the breath, while arnica alleviated bruises. Women also made their own household products; rosemary made an excellent hairwash, and dried lavender deterred clothes-moths.

Herbs and spices were also vital for cooking; herb gardens generally contained chives, parsley, rosemary, mint, sage, thyme, fennel, bay, caper and tarragon. In an era before refrigeration, the cook often had to disguise over-ripe meat or fish that was past its best; a soak in milk containing charcoal, a quick rinse and an herb-rich sauce could conceal the awful truth.

ABOVE: A stone bowl of thyme in the herb garden at Sissinghurst Castle Garden, Kent.

BELOW: Chives growing in the kitchen garden at Biddulph Grange, Staffordshire.

ABOVE: A wigwam of sweet peas has been planted in the middle of a bed of vegetables.

BELOW: Carrots are planted in rows with marigolds interspersed to deter pests.

beech tree in the middle to provide shade. Visitors will be encouraged to occupy the seats and arbours while they appreciate the kitchen garden.

The plantings

The plants initially grown in the open beds include courgettes, cabbages, carrots, spring onions, calabrese, spinach, potatoes and chard, with a wigwam of sweet peas in the middle of a bed of cauliflower and cabbage. Runner beans climb up a cone of salvaged branches in the centre of another bed. In a witty touch, the letters 'VR' have been planted in royal purple cabbages surrounded by calendula, a horticultural homage to Queen Victoria.

Where possible, Victorian varieties have been chosen. Many 'heritage' vegetables have distinctive flavours, such as the 'Kelvedon Wonder' peas, but they need to be resilient too so the selection has been made with care, mixing the best of the old with the most appropriate of the new. In addition, an assortment of culinary herbs is being cultivated; herbs were essential to add zest to Victorian cuisine.

The glasshouse and the potting shed

Every Victorian gardener of any ambition wanted a decent-sized heated glasshouse so that he could grow exotics for his employer. New methods of manufacturing sheet glass and supporting the panes in a skeletal framework of wrought iron offered infinite possibilities to the advanced gardener, and the removal in 1845 of the exorbitant tax on glass helped to promote the possibilities of this essential adjunct to any kitchen garden. The epitome of the Victorian glasshouse was created by Joseph Paxton, gardener to the Duke of Devonshire, whose visionary expertise enabled him to create the Crystal Palace – essentially a giant pre-fabricated greenhouse – to house the Great Exhibition of 1851.

RIGHT: The handsome new glasshouse has been erected against the eastern wall of the kitchen garden, using some of the original brickwork from the glasshouse that was there in the nineteenth century.

THE VICTORIAN GARDENER

Nineteenth-century gardeners on country estates were recruited as teenagers, with sons often following in their father's footsteps. They learnt their skills by instruction from their superiors and practical application, eventually rising to the position of head gardener if they were fortunate. Professional gardeners rarely recorded their technical knowledge, preferring to keep their own counsel so as to keep their jobs.

Gardeners on large estates needed to be good all-rounders, able to turn their hands to coppicing and track-laying, as well as all kinds of horticulture, topiary, crop rotation and kitchen garden produce. The Victorian head gardener was also responsible for starting off, nurturing and transplanting to the formal gardens a huge variety of bedding plants, as well as decorative pot plants such as orchids to adorn the interior of the house.

His most important role, however, was to maintain a constant supply of fresh fruit, vegetables and cut flowers for the big house throughout the year, a task that required considerable ingenuity,

BELOW: The gardener at Avebury in the nineteenth century during the Kemms' tenancy. The garden looks very well tended.

especially over winter. Every morning, the head gardener would consult the cook as to the decisions that had already been made for that day's meals. He would advise on the fruit and vegetables that were in season, and a selection of the best produce would subsequently be cut or picked and delivered to the back door of the kitchen.

Some householders were extremely picky; Mrs Greville of Polesden Lacey in Surrey, for example, insisted that fresh fruit should be supplied from the garden to the kitchen in decorative baskets lined with leaves, and she would not tolerate any misshapen produce. Fruit and vegetables that were less than perfect were either consumed by the servants, or despatched to the local greengrocer in nearby Bookham for sale to the less discerning general public.

ABOVE: Fresh fruit and vegetables from the kitchen garden at Llanerchaeron, Ceredigion. The best of the seasonal produce was delivered directly to the back door of the kitchen.

THE WILTSHIRE GIANT

Fred Kempster, the 'Wiltshire Giant', trained as a gardener at Avebury Manor before becoming an Edwardian circus attraction. Born in London in 1889, he was taken to Avebury to live at the age of four. He suffered from gigantism, caused by the over-production of a growth hormone, and his hands alone measured 38cm (15in) from fingertip to wrist! When he reached the height of 2.1m (7ft) he joined a travelling circus, where he was billed as The World's Tallest Man; the public were fascinated by physical oddity, and mild-mannered Fred was considered a freak. He continued to grow, eventually reaching the height of 2.44m (8ft 4in). The circus was in Germany when the First World War broke out and the performers were imprisoned in terrible conditions. On his release in 1916, Fred returned to Britain but succumbed to the great flu epidemic in 1918, dying at the age of 29.

Victorian glasshouses were used to grow delicacies; tomatoes were considered an exotic fruit, rather than a staple of every salad, as nowadays. Aubergines, melons, grapes, peaches and pomegranates were nurtured, as well as pot plants and cut flowers for the house. The gardeners might cultivate chrysanthemums for Christmas, bring on lily-of-the-valley to fill the house with fragrance over New Year, or produce hundreds of highly scented hyacinths for a party.

At Avebury Manor, a handsome glasshouse has been erected against the eastern wall of the kitchen garden, where the brickwork still bore visible traces of the substantial structure which had been there in the nineteenth century. The new glasshouse is modern, an ex-display model supplied by specialist firm Alitex; it proved impossible to locate a Victorian glasshouse robust enough to be dismantled and re-erected on site. The three-quarter span glass roof maximises the effects of any available sun from all directions, all day. It is fitted with vents at roof level to allow natural ventilation and so reduce the risk of fungal disease.

A venerable potting shed has been bought from a nearby location and re-erected on site. Potting sheds were an essential feature in any Victorian

NEWT MANOR

While the glasshouse was being erected in the kitchen garden at Avebury, the workers were excited to discover a pair of great crested newts (*Titurus cristatus*). These rare amphibians are a European Protected Species, so it was essential to protect their habitat. They were therefore provided with a secure little structure of their own down by the compost bins, which was soon nicknamed 'Newt Manor' by the staff and volunteers. Specialist painter Mark Sands even included a picture of the newt in his exotic wallpainting in the Queen Anne Antechamber.

garden; they offered the gardeners a refuge from poor weather, a workplace for essential backroom activities such as preparing cuttings and sowing seeds, and a place to store all the tools and equipment from dibbers to rakes, earthenware pots and secateurs.

Green gardening

David Howard acknowledges that horticulture subverts what would occur in nature; seeds normally sprout wherever they have an opportunity, regardless of whether the conditions are ideal for that sort of plant, but this process is managed and controlled in a garden. However, he believes in applying the principles of organic gardening to every setting; by intelligent analysis of a given site, problems can be largely overcome using natural remedies rather than by dousing plants with herbicides or pesticides. In the Victorian era, sprays of nicotine solution over tender fruit and vegetables certainly deterred insects, but they might also be injurious to human health. Employing organic methods of gardening removes the need to use harsh chemicals.

ABOVE: Tools in the potting shed.

BELOW: Fee Robinson (left) of the Kennet Bee Keepers Association shows Penelope Keith (right) a beehive at a cottage in Alton Priors, a small village near Avebury.

'Companion planting' is an organic method of deterring insects by growing a particular plant close to the one that needs protection; for example, siting pungent marigolds alongside carrots masks the latters' scent, by which the carrot fly would otherwise find the host plant. Other strategies include planting later in the season to avoid predatory insects at crucial stages of their development. Elsewhere, David has previously employed the ingenious strategy of introducing chickens into a client's orchard to eat the over-wintering pests; the fowl's manure helped fertilise the soil, improving the yield of the fruit trees, which in turn supported a number of beehives, and the bees then pollinated other plants in the area. As a result, the orchard produced not only bumper crops of fruit, but honey and free-range eggs as well.

Of course, some insects are beneficial as they prey on pests or help to pollinate plants. At Avebury Manor, bees are encouraged by the deliberate planting of fragrant blooms such as sweet peas and lavender. In addition, a working hive has been put in the kitchen garden, positioned behind the rabbit fence at the end of the vegetable beds, out of reach of inquisitive visitors.

ABOVE: Penelope Keith prepares to cut the ribbon at the grand opening of the kitchen garden. With her are David Howard and Paul Martin (on the left) and Jan Tomlin and Sarah Staniforth of the National Trust (right).

RIGHT: The tools of the gardeners' trade ready for use.

OVERLEAF: The finished kitchen garden in all its splendour.

Compost and recycling

In the nineteenth century, many progressive estate owners bought tonnes of guano – bird excrement rich in nitrogen and phosphates – to add to the soil as a fertiliser. It was shipped to Britain from South America, and there were fortunes to be made from importing this malodorous cargo; the Gibbs family was able to establish a magnificent house and estate at Tyntesfield near Bristol on the proceeds of their trade in Peruvian guano. However, there is no need to use foreign materials at Avebury; by the careful husbanding by the gardeners of all suitable material generated on the estate, including leaf litter, lawn clippings and straw, the kitchen garden should be able to generate its own compost to improve the soil. To this end, compost bins have been constructed in one corner of the kitchen garden, using recycled wooden pallets.

Traditional Nettle Beer

Little went to waste in Victorian times and country folk would have made a variety of drinks from whatever was growing locally, such as dandelion and burdock or elderflower cordial. The inhabitants of Avebury may well have brewed nettle beer – a refreshing, mildly alcoholic beverage popular throughout Wiltshire – which is easy to make.

Recipe

Wearing rubber gloves, pick about 900g (2lb) of young nettle tops. Rinse them thoroughly, then boil them in 4.5 litres (1gal) of water for 30 minutes. Strain the liquid, discard the leaves and dissolve 200g (7oz) sugar in the liquid by stirring. Add 1 1/2 heaped teaspoons of ground ginger and pour the mixture into a large sterile container.

Spread some fresh yeast on a small slice of toasted bread, and add it to the liquid – it will float. Cover and leave for 3 days, then strain and decant the liquid into clean screw-top beer bottles. It will be ready to drink after 2 days; handle with care, as it can be very fizzy.

CHAPTER V

AVEBURY IN THE EARLY 20TH CENTURY

EARLY IN THE NEW CENTURY, AVEBURY MANOR PASSED INTO THE HANDS OF A COUPLE WITH A PASSION FOR RESTORATION AND REFURBISHMENT. THE JENNERS INTRODUCED ANCIENT OAK PANELLING AND ANTIQUE FURNITURE WHERE NONE HAD EXISTED BEFORE, WHILE THEIR SUCCESSOR, THE MILLIONAIRE ALEXANDER KEILLER, COMBINED A SPARTAN ENTHUSIASM FOR THE TUDOR ERA WITH A LOVE OF THE TRAPPINGS OF THE MACHINE AGE.

LIFE IN THE MODERN ERA

The death of Queen Victoria in 1901 and the succession of Edward VII, an altogether more pleasure-loving monarch, ushered in a period known as the 'Edwardian summer'. However, this era was less idyllic than it appeared on the surface. Considerable social tensions existed at home with calls for workers' rights and female emancipation. In 1914 European conflict sparked the first fully mechanised war, during which few homes escaped the visit of the telegram boy bearing a few brief lines signalling the end of hope. Rationing and taxes, inflation and death duties threatened many households and even aristocratic estates were abandoned or sold at rock-bottom prices.

The carnage and its end were commemorated in Sir Edwin Lutyens' dignified design for the Cenotaph in

LEFT: A First World War recruitment poster, from about 1917, urges the 'Women of Britain' to encourage their loved ones to go to war.

EARLY 20TH CENTURY TIMELINE

1901
Edward VII succeeds Queen Victoria, heralding the 'Edwardian summer'

1914–18
Britain is plunged into war in Europe; the Great War is thought to be 'the war to end all wars'

1923
The Charleston is performed in the hit Broadway show *Runnin' Wild;* it soon becomes a worldwide dance craze

1910
George V accedes to the throne

1918
As the troops return home, Prime Minister Lloyd George calls for 'a land fit for heroes'

Whitehall, unveiled in 1920. With the hardships of the war years left behind, hedonism took over in the following decade, which was characterised by a frenetic enthusiasm for novelty and noise in the form of gramophone records, cinema, radio, telephones, parties, confrontational art and foreign travel. The 'Roaring Twenties' were also a time of a burgeoning British infatuation with America, the home of professional entertainment and wealthy heiresses who could replenish the family coffers.

ABOVE: An illustration entitled *Paris by Night* (*c*.1925), by Edmond Haraucourt, perfectly captures the hedonistic mood of the 'Roaring Twenties'.

Edward VIII

1926
The General Strike causes the nation to grind to a halt for nine days

1936
George V is succeeded by his son, Edward VIII, who abdicates soon after; Edward's brother becomes George VI

1925
Paris holds an international exhibition of *arts décoratifs*, epitomizing a style later known as Art Deco

1929
The Wall Street Crash sees the value of shares on the New York Stock Exchange plummet, heralding a worldwide depression

1939
The Second World War begins and the nation prepares for the fight against fascism

The return to war

The fun was brought to a close by the Wall Street Crash of 1929, which ushered in a worldwide economic depression. Throughout the troubled 1930s, financial instability threatened the great powers, while dictators flexed their muscles in Europe, Russia and the Far East.

Perhaps in reaction to the uncertainty of the present, there was a growing fascination with Britain's past; at Avebury, Alexander Keiller used his own fortune to excavate the largest Neolithic site in Europe, while at Sutton Hoo the archaeologist Basil Brown uncovered a unique Saxon burial site. But meanwhile, trouble was brewing overseas. Many Britons had thought that Herr Hitler was 'a good thing', a credible leader bringing stability to a financially hard-pressed and humiliated country, and that Germany would never repeat the devastation of two decades before. They were wrong; Europe slid once again into war, this time one that lasted six long years during which the majority of the populace were either on active service or taking part in difficult and often dangerous duties on the Home Front. When VE Day finally arrived in 1945, the British were exuberant but utterly exhausted.

House and Home in the Twentieth Century

New domestic buildings in the first quarter of the century paid lip service to the vernacular vocabulary made popular by Charles Voysey and Philip Webb, William Morris's favourite architect. Black and white half-timbering, pitched overhanging gables and decorative panels of stained glass, depicting galleons at sea or bucolic country scenes, reflected the nostalgic reverence for 'Merrie England', a supposedly idyllic era before industrial mass manufacture and mechanised destruction.

After the Great War, there was a drive to create 'homes fit for heroes', and a corresponding boom in speculative housing; the semi-detached suburbs of 'Metroland' spread out from the cities. Crucially, the middle classes were also learning to exist independently of servants for the first time. A 'daily' might tackle the muckier tasks, but electrical gadgets such as affordable domestic vacuum cleaners now replaced the daily round of live-in staff.

For all their compact modernity, the interiors of these revisionist houses reveal a love of the past; Classicism and Regency style found a new appreciation, while the Tudor Revival ushered in a demand for reproduction furniture, often made of carved and turned oak, fumed to give the impression of age. Upholstery textiles tended to be dark and textured with traditional patterns such as fleur-de-lis, and oak wainscoting adorned many a suburban sitting room.

It was as though a certain section of the population hankered after a historic past. They might take up smoking and drink gin

ABOVE: This poster from *c.*1910 depicts a Tudoresque suburban home, complete with timber-framed gable and leaded windows.

INNOVATION IN THE HOME

As the supply of electricity to homes became increasingly widespread throughout the 1920s and '30s, a growing number of people were able to take advantage of the plethora of innovative electrical devices that transformed everyday tasks from cleaning and ironing to cooking and keeping food fresh. Many of these had their roots in designs dating back to the Victorian era but only became practicable in the early decades of the twentieth century. For example, while the first patent for a fridge which relied on a compressed gas, or refrigerant, for its cooling effect was filed in 1851, electrically powered domestic models were not produced until 1913.

Other popular innovations included the electric iron (dating back to 1882), the electric kettle (the first efficient model being made by the Swan Company in 1921), the electric hairdryer (first manufactured around 1920), and the domestic pop-up toaster (from 1926). Possibly the most labour-saving device of all was the electric vacuum cleaner. The first powered model, driven by petrol engine, had been invented in 1901 by Herbert Booth, but it was so big it had to remain in the street while a long hose was fed into the house to carry out the cleaning. The first electric model was brought out by Hoover in 1908, and before long vacuum cleaners had become a mass-market product.

ABOVE: An advertisement for vacuum cleaners, from 1924, emphasises the 'wizardry' of owning one of these appliances.

LEFT: This magazine illustration from the 1930s shows a modern housewife surrounded by some of the latest kitchen innovations.

gimlets, dance the Black Bottom to gramophone records and buy 'Bizarre' Art Deco tea-sets, but when it came to domestic bricks and mortar, they preferred the solidity of Revivalism to the more transitory uncertainty of the Jazz Age. Flat-roofed Modernist buildings were acceptable as lidos or picture palaces, and streamlined concrete curves evoked the efficiency of the newly built suburban stations, but they were too novel and 'foreign' for most home-builders, who preferred to cling to the styles of a safer, less treacherous past.

The conservation craze

This uniquely British attitude towards the past grew from a genuine groundswell of enthusiasm for heritage conservation. The first Ancient Monuments Protection Act had been instigated in 1882 by John Lubbock MP, who had purchased large parts of Avebury in 1871 to protect it from exploitation; significantly, when he was ennobled in 1900 he chose the title of 1st Baron Avebury. The Society for the Protection of Ancient Buildings (SPAB), which counted William Morris among its founders, campaigned for the care of threatened structures, and the National Trust, founded in 1895 to look after places of historic interest and natural beauty, in the 1930s started to acquire important country houses – previously the domain of the rich.

The passion for conservation was reflected in *Country Life* magazine, which was launched in 1897 and read by both country and urban readers. From its inception, the magazine photographed and described venerable private houses, and by 1913 was described as 'keeper of the architectural conscience of the nation'. Those sympathetic to the cause were known as the *Country Life* Circle.

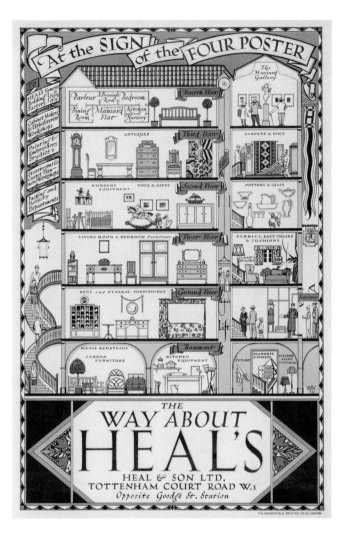

ABOVE: An advertisement for Heal's furniture store, *c.*1920, reflects the popularity of traditional designs that hark back to the past, as opposed to the Modernist styles of the future.

CASTLE DROGO

ABOVE: For the main staircase, with its domed and vaulted ceiling and huge mullioned windows, Lutyens created a modern interpretation of a Norman castle.

BELOW: In contrast to the staircase, the drawing room at Castle Drogo was inspired by eighteenth-century interiors.

While heritage adherents, such as those belonging to the *Country Life* Circle, were restoring old houses in period style, some architects were creating 'old-style' houses from scratch. In 1902 Edwin Lutyens was commissioned by Edward Hudson, publisher of *Country Life*, to build a home on the ruins of a castle on Lindisfarne Island. The result was an architectural triumph and attracted the attention of the grocery millionaire Julius Drewe, who was determined to prove his family link with a Norman noble, called Drogo or Dru, who had once held substantial estates in Devon.

Drewe desired a vast fortress overlooking Dartmoor, and in response Lutyens designed a sophisticated interpretation of various historic styles with Norman, medieval, Tudor and eighteenth-century references, implying that the castle had developed over the centuries. Nevertheless, Castle Drogo was technologically advanced, with central heating and power from a hydro-electric plant in the gorge below. Neither architect nor patron could have imagined that, within three years of work commencing in 1911, the skilled workforce would be fighting for their lives in the mud of Flanders. Not only the labourers were cut down; Drewe's eldest son, Adrian, was killed in battle in 1917. After the Great War, the demise of the low-paid servant classes meant that Lutyens' masterpiece was to be the last castle built in England.

Avebury in an Age of Conservation

Leopold and Nora Jenner, who moved into Avebury in 1902, were part of a group of well-heeled people who were interested in acquiring old houses in order to restore them. Avebury Manor and Lytes Cary, the latter a medieval and Tudor manor house near Ilchester in Somerset, were both on a SPAB (Society for the Protection of Ancient Buildings) list of buildings at risk when they were purchased in 1907 by Leopold and his brother Walter Jenner respectively.

The Jenner brothers co-operated in sharing craftsmen and specialist restorers to work on their historic houses.

Walter's architect at Lytes Cary was C.L. Ponting, based in Marlborough and known for his restoration work on old stone churches. He was engaged to design the new wing at Lytes Cary, and it is possible that he was also the architect of the Jenners' new library at Avebury, as the brothers co-operated in sharing craftsmen and specialist restorers to work on their historic houses. Both brothers patronised a Mr Angell of Bath, who sold and restored antique furniture. A skilled craftsman, Angell made a replica doorway for the dining room at Lytes Cary to match the one reclaimed from a demolished Wren church in London. He also installed various reclaimed materials and fixtures for the

ABOVE: Leopold's brother, Sir Walter Jenner, rescued Lytes Cary from dereliction and started to restore it, furnishing it with authentic seventeenth- and early eighteenth-century oak pieces.

Avebury Jenners, who, as early proponents of architectural salvage, had seventeenth-century panelling fitted in some of the rooms. In others, such as the old library (now the Billiard Room) the panelling was specially made in a seventeenth-century style.

THE MANOR RESTORED

The Jenners lovingly restored Avebury Manor and furnished it in a manner appropriate to its long history. There had been a two-day sale of all the contents in 1902 when the Kemm family relinquished the lease, so the house was virtually empty when the Jenners moved in as tenants; it was also very dilapidated,

Nora and Leopold undertook essential repairs; the drawing room had huge cracks in the walls, and one wall had to be rebuilt, a process that took nearly two years. The chimneypiece, which belonged in the Tudor Parlour, had been removed in the nineteenth century, but its damaged components were discovered in 1907 in an outbuilding and it was repaired and reinstated. To tackle the perennial problem of cold, a solid fuel central heating system was installed, though with the flawless timing of all household appliances this broke down in January 1915 at a time when the household was laid low with flu.

ABOVE: The Jenners reinstated the original Tudor fireplace in the parlour, after discovering it in an outbuilding.

LEFT: Unlike Avebury Manor, Lytes Cary has a full-height Great Hall. The open timber roof has carved angels on the ends of the rafters.

BELOW: A photograph from the Jenner period shows the recently installed panelling in the South Library at Avebury.

ABOVE: Commissioned by the Jenners, the iron balustrade on the external staircase leading up to the new library has all the delicacy of an Adam's style arabesque.

The restoration was necessarily a slow process. Shortly before Leopold rejoined the Army in 1914, ironwork had been commissioned and installed in the grounds, including new gates and screens, but in places the gardens were nothing more than heaps of earth. Nevertheless, the Jenners' enthusiasm for Avebury Manor did not dim; they even used the existing structure as inspiration for new developments, echoing the pattern on the Tudor Parlour's ceiling in the geometric hedges in the topiary garden, which they added in 1921. They also delighted in furnishing the house with appropriate art and artefacts, scouring the country for suitable antiques. A consummate needlewoman, Nora even embroidered the voluminous bed hangings for the Queen Anne Bedroom with great skill and attention to historical techniques and materials, visiting the textile collection of the Victoria and Albert Museum for research.

A new owner at Avebury

Financial worries and the difficulty of retaining servants finally forced the Jenners to retrench and they left their beloved Avebury Manor in 1929 to live in Bath. They leased the house to the millionaire archaeologist Alexander Keiller in 1935 and two years later he bought the manor and all its associated buildings, having already acquired large tracts of the surrounding landscape. Parts of the house that had fallen into disrepair were restored and Colonel Jenner's library was filled with Keiller's enormous collection of books on every subject from anthropology to witchcraft.

Though he was very much a man of the Machine Age, with a restless enthusiasm for speed, gadgetry and new technology, Keiller retained a reverence for antiques and time-honoured traditions. A contemporary noted that everything in the house was brown – curtains, carpets, window seats, leather-covered walls – in keeping with prevailing 1930s fashions. Keiller's bedroom and dressing room on the top floor were quite Spartan, with bare floorboards covered with animal skin rugs. Keiller himself described these private rooms as 'equipped with that indescribable degree of personal discomfort which can only come from exclusively period furniture of the middle of the XVI century'.

In 1943, Keiller sold 384ha (950 acres) of the Avebury estate to the National Trust for £12,000 – a sum worth about £311,400 in modern terms – though he retained the manor until 1955 when he sold it to the scientist Sir Frances Knowles. Throughout the last century, Avebury Manor was undoubtedly saved from probable disintegration or deliberate demolition by a succession of dedicated individuals who had the energy, the finances and the foresight to ensure its future.

ABOVE: A contemporary photograph of his dressing room shows the 'degree of personal discomfort' that Keiller put up with in his private apartment.

TOURISM COMES TO AVEBURY

Fuelled by the press coverage of Keiller's archaeological excavations, tourists started to visit Avebury in increasing numbers. The Red Lion Hotel offered accommodation and catering for parties of up to 200, as well as the hire of buses and a 'Motor Landaulette' to pick up guests at Marlborough or Swindon stations. Meanwhile, train companies and other interested parties were encouraging leisure travel by producing avant-garde posters such as this iconic one of 1934, by artist Edward McKnight Kauffer, for the Shell Petroleum Company.

THE GOLDEN AGE
OF MOTORING

Both the Jenners and Alexander Keiller were enthusiastic car owners; Nora even drove her own automobile before the First World War – a rarity for a woman in those days. The earliest motorcars enabled a few wealthy people to travel further and faster than ever before in a private vehicle, albeit in conditions of some danger and discomfort. The future Edward VII, an inveterate thrill-seeker with a low boredom threshold, took his first automobile ride in 1896 and acquired his first motor in 1900. Society figures rapidly followed his example, initially driven by chauffeurs but gradually learning to drive themselves.

Petrol stations were almost unknown; instead, motorists would buy two gallons of petrol in metal cans from their local chemists. Ladies swathed themselves in 'duster coats' and full-face veils, while gentlemen donned goggles, tight-fitting leather caps and gauntlets. In 1905, there were 8,000 motors on the roads of Britain; by 1914 this figure had swelled to 132,000.

The first affordable British cars were produced by Morris and Austin after the First World War. The popular Austin Seven went into production in 1922 and had sold over half a million by the time it ceased manufacture

ABOVE: Back in 1932, as now, car manufacturers sold a lifestyle. For a cost of £159, the Hillman Minx promised two stylish young women the freedom of the open road.

in 1939. The smallest of these was the Chummy, a tiny open tourer, seating two adults, with a maximum speed of 60km/h (38mph). More sober types preferred the Austin saloon which could take three children in the back and offered some protection from inclement weather. At Avebury, the internal combustion engine brought other benefits; motor buses now linked the village with bigger towns such as Devizes and Marlborough.

RIGHT: The very wealthy might own a fleet of cars run by a small army of chauffeurs and mechanics. This fleet, dating from 1928, belonged to Mrs Greville of Polesden Lacey in Surrey.

KEILLER AND HIS CARS

In the museum at Avebury stands a magnificent car owned by Keiller from 1914 until his death in 1955. This Anglo-French rarity (right) is a 20hp Malvern Torpedo Tourer with a 4-cylinder 4060cc engine, made by Sizaire Berwick; unusually, the front seat converts into a double bed. Keiller helped to finance the Sizaire Berwick company, and he acquired the precise drafting skills so valuable to an archaeologist by spending hours in the company's design studio.

Keiller owned several other notable cars, too, including a famous Hispano-Suiza with tulipwood coachwork. In 1929, he and a Miss Duncan barely survived a 135km/h (84mph) crash in his Targa Florio Bugatti. The back axle broke when the car hit a bridge, and the occupants nearly plunged over the balustrade onto a railway line 12m (40ft) below. Keiller also ran a Citroën Kégresse half-track – an open estate car with tank tracks instead of rear wheels – which he used on his archaeological excavations; it was destroyed in a fire in 1945.

LEOPOLD AND NORA JENNER

In 1902, Colonel Leopold and Nora Jenner, along with Nora's daughter Ruby, aged nine, moved into Avebury Manor. They adored the house and their lives in that Edwardian summer were idyllic. After renting for five years they bought the manor in 1907 and embarked on an ambitious programme to restore it to the way it might have looked throughout its venerable history.

Leopold Jenner was the son of the indefatigable and wealthy William Jenner, who, as physician and close personal friend to Queen Victoria, was the most eminent medical man of his generation. Leopold joined the army at the age of 18 and served in the First World War with distinction in Egypt, Gallipoli and France. His military career, however, was overshadowed by that of his remarkable father-in-law, Field Marshall Sir Donald Stewart, a national hero who had been Commander-in-Chief of the British Army in India. In recognition of his standing, Sir Donald received a state funeral upon his death in 1900, with a personal wreath from Queen Victoria resting on his coffin.

With his bushy handlebar moustache and military manners, Sir Donald had been a distant, intimidating figure to his three young daughters, and there was a savage rift when Nora, aged 23, married

BELOW: Nora Jenner (at the table) sits in the garden at Lytes Cary, Somerset, with members of her family.

without his permission. In 1894 the marriage ended after only three years in a sensational divorce which was even reported in *The Times*, but in 1899 Nora met Leopold Jenner and their subsequent marriage was happy and long. Two of the sisters, Flora and Nora, remained very close; Flora married Leopold's older brother Walter, who bought and restored the ancient manor house of Lytes Cary, so the two Jenner families lived near each other and shared their passion for conserving old buildings.

Life at Avebury

Leopold, a keen polo player, was a respected figure in the village, often seen in his Bentley or taking the air with his two Dalmatians. Nora drove her own small car; an emancipated woman who had survived a domineering father and a very public divorce, she was now living life on her own terms. She was certainly aware of the Women's Suffrage Movement; in a letter to her sister Flora describes how she had once been made to leave her handbag in the cloakroom while visiting Haddon Hall, a Derbyshire country house, 'on account of the suffragettes'. The stipulation was doubtless in response to protest attacks by militant suffragists, who concealed small rock hammers in their handbags with the intention of damaging nude male statues in London galleries.

ABOVE: Leopold Jenner was a keen polo player, seen here on his horse, Lorna Doone.

Life in Avebury was convivial; there was an annual fête and the village children were invited to the manor for fruit cake and lemonade. One year, Nora Jenner organised a Christmas pantomime at the Manor, painting the sets herself. She also worked with the other ladies of the village to sew the actors' costumes. Both sisters were expert needlewomen; Nora embroidered many of the textiles for Avebury, including the curtains and covering for the marital bed, and also revived the art of stumpwork, a seventeenth-century needlework technique which gives a three-dimensional effect to flat textiles.

ABOVE: A photograph dating from the 1920s shows Avebury Manor as seen through the east entrance gate, with the old stable building on the right.

According to the 1911 census there were seven staff at Avebury Manor; two ladies maids, two housemaids, a cook, kitchen maid, and parlour maid. The Jenners were well-connected and it is thought that Nora knew Vita Sackville-West, who visited Avebury during the 1920s and was said to have admired the Jenners' garden. Towards the end of their time at the house, Colonel Jenner decided to have the Manor wired for electricity – Mrs Jenner was reputed to be tired of peering at her ancestors on the dining room walls by candlelight.

Hard times

The Jenners had been comparatively wealthy; Leopold had inherited a share of his father's impressive estate of £300,000 in 1898, and Nora had an inheritance from her mother, though this was held in trust. Money became a problem for the couple in the aftermath of the Great War, when they suffered from diminished incomes, increased taxes and exorbitant prices. Inflation was also a headache; in 1914, £1 was worth approximately £90 in today's money, but by the end of the war its actual purchasing power had dropped by half. Nora also worried about the shortage of servants. It was difficult to attract and retain staff in the country because they preferred the attractions of city life and, without hired help, life in Avebury Manor was definitely less comfortable.

In 1929, the year of the Wall Street Crash, the Jenners rented out the Manor after 27 years of life in Avebury and moved to Bath. Anecdotal evidence suggests that losing money in Chilean Railways was the final straw for them financially. Yet they chose to be buried in the Avebury churchyard. Nora predeceased Leopold, who wrote a simple but profound epitaph for her: 'The most perfect Wife and companion a man ever had.'

BELOW: Although the Jenners left Avebury Manor at the end of the 1920s, they both chose to be buried in the graveyard of Avebury church.

THE BILLIARD ROOM

This atmospheric room lies in the west end of the south wing, in a part of the house that dates from about 1600, but it is not as it would have been in the seventeenth century. In a sketch of Avebury dating from 1695 it is shown divided into two, forming part of the service area of the house and containing a pantry, for the storage of dry foodstuffs, and a room marked as the 'inner cellar'. A recent survey of the manor house also suggests that these rooms may have had very low ceilings, with a extra floor squeezed in above them, and lying below the bedchamber (now the Queen Anne Withdrawing Room) on the top floor.

This room is not as it would have been in the seventeenth century. In a sketch dating from 1695 it is shown divided into two, forming part of the service area.

According to a report in *English Homes*, a book published in 1929 by *Country Life*, when the Jenners took over the lease to Avebury in 1902 they apparently found this space 'unfloored, disused, almost ruinous'. They subsequently spent time and effort transforming this largely derelict area into a library in keeping with the style prevailing at the end of the seventeenth century, when Sir Richard Holford owned Avebury. They installed robust, large-format wood panelling, specially crafted for the room, and a moulded marble fire-surround similar in style to those fitted at Hampton Court during William III's reign (1688–1702). The fireplace is made from maroon and green marble of a type known as Cornish Serpentine stone, and has deeply curved stone carving, or bolection moulding, which was also popular around 1700. Photographs of the room in the 1920s reveal decorative lidded Chinese jars on the sills of the mullioned window and a liberal assortment of books.

The South Library, as it was known, was obviously a comfortable room, even if it was not true to its original purpose, for there are very few bookcases here. Sir Francis Knowles, a later owner of Avebury Manor, commented of the Jenners' restorations that 'where they could replace the old fabric they did so, and where they could not they imported woods of suitable date from other old houses … it has been said of the Jenners that they and Avebury Manor were perfectly suited to each other … they left Avebury Manor what it is today – a home where taste has been blended with comfort and tradition'.

BELOW: The fireplace installed by the Jenners has robust bolection moulding and dates from around 1700.

Transforming the Billiard Room

To reinterpret this room in a way that represented the Jenners' life at the house immediately after the First World War, designer Russell Sage and the experts identified a number of key themes based on extensive historical research. It was an overtly masculine room and served as Leopold Jenner's specially created 'den',

a place where he could relax undisturbed with his books, or invite friends for conversation or sport. He was a military hero who on the outbreak of the Great War had insisted on rejoining the army at the advanced age of 45, serving in Egypt, Gallipoli and France. A fit and active man, in his youth Jenner excelled at a number of sports, winning the Army Fencing Championship in 1894–5, and on his first retirement from the army he became joint Polo Manager of the Ranelagh Club (1904–11). He played in many prestigious polo tournaments and was selected to play for England versus Ireland in 1907, so it was appropriate to highlight his love of all sorts of competitive sports, his military record and his other interests as a huntsman.

Russell and the experts felt that this room should represent a gentleman's preoccupations and leisure pursuits at a point just after the Great War, so they decided to turn it into a billiard room. It was agreed that a scheme of deep reds combined with the dark wooden walls and the green baize would be most appropriate here, with a number of military mementoes and sporting trophies to emphasize the masculine ambience of an environment designed for comfort, recreation and reflection.

Trophies on Display

Stuffed animals, mounted on the walls as trophies or displayed in glass cases, were a feature of British country houses before the Great War, as both scientific exploration and game-shooting were seen as gentlemanly pursuits. A taxidermist would remove the skin or pelt and preserve it, then create an intricate wire, wood and plaster armature. Gaps between the pelt and the 'skeleton' underneath it were stuffed to create a realistic replica of the creature in life. Glass eyes, false teeth and plaster tongues were added, and the exhibit was treated with arsenic to deter insects. At Kedleston Hall, in Derbyshire, the Trophy Corridor (right) reflects the popularity of this practice.

The Billiard Table

Dominating the room is a very substantial six-legged mahogany billiard table, made by the firm of George Edwards of London some time between 1870 and 1925. The specialist firm of Hubble Sports, established in 1910, who are experienced in refurbishing the more unusual billiard tables and undertaking authentic restorations, purchased this table from a family in Kent who had owned and used it for many years. The firm completely restored the piece in their workshop on the Leeds Castle estate, replacing and shaping the cushion rubbers and resurfacing the table with new baize made from pure English wool from Pudsey in Yorkshire. They gave the frame a traditional French polish, a lengthy and arduous process, using their own recipe, which blends with the existing polish.

The table has a five-piece Welsh slate bed and consequently weighs a hefty 1200kg (1⅓ tons). The accompanying accessories include a handsome billiards scoreboard and a cue-rack. Above the table, throwing its light onto the surface, is an electric chandelier, suitably shaded in green, to make the billiard table the focus of the room.

ABOVE: Peter Ludgate of Hubble Sports fits the heavy Welsh slate bed onto the billiard table *in situ*.

A Game Fit for a Queen

The game of billiards has been played in Britain since at least the sixteenth century, though its origins remain obscure. Mary, Queen of Scots complained vociferously in 1587 that her table had been taken from her during her captivity at Fotheringhay Castle and Shakespeare referred to billiards in *Antony and Cleopatra*, first performed in 1606–7. The earliest known set of rules in English was published in 1650, and in 1679 a billiard table is known to have been set up in the hall of Ham House by the Duchess of Lauderdale for the amusement of her guests.

Competitive games of skill that could be played indoors in inclement weather were considered essential in British country houses. Queen Victoria had a billiard table installed at Windsor Castle in 1838 and, between 1835 and 1870, more than two-thirds of new country houses had a purpose-built billiard room on the ground floor, often with a smoking room, a gun-room or study attached. This suite formed the 'bachelors' wing', with bedrooms for male guests above. Gentlemen attending

BELOW: The Billiard Room at Polesden Lacey, Surrey, was designed by Ambrose Poynter and resembled a gentleman's club.

houseparties would play billiards in the afternoon, between lunch and dinner. After dinner they had social obligations to maintain but on less formal evenings, when there were no guests, male members of the family would be likely to practise their shots once the meal was over.

The tables were extremely heavy as the beds were made of polished slate placed on an oak frame then covered in green baize. A wooden cue rack and a scoreboard mounted on the wall completed the necessary equipment. The purpose-built Billiard Room at Polesden Lacey was created as a gentleman's club in miniature, a plush bastion of masculinity and a venue for men to talk business, politics and sport. Its counterpart at Tyntesfield is a more spartan affair, festooned with antlered sporting trophies, though the table itself is heated by hot water pipes, a great boon in an otherwise chilly room.

In Queen Victoria's day billiards was largely a male preserve, but by the Edwardian era there were instances of female society hostesses prepared to defy social convention and wield a cue. Canny advertisers promoted the game as a suitable hobby for all the family to play together. There was even a movement to open Temperance Billiard Halls, offering young men wholesome alternatives to spending evenings in the pub.

ABOVE: The Billiard Room at Castle Drogo, Devon, was designed by Sir Edwin Lutyens and was decorated to look like a medieval chamber.

BELOW: By the beginning of the twentieth century billiards had become a game widely enjoyed by women.

The Finishing Touches

Other furniture was brought in to add to the cosy air – none of it especially noteworthy, but all 'good country house pieces', to use the auctioneers' jargon, and very much what the embattled British gentry would pick up at auctions and sales. Two upholstered wing armchairs, each flanked by a side table for setting down drinks and books, are arranged to make it 'a delightful room for converse or study', according to *English Homes*. A pair of semi-circular tables set on either side of the fireplace support reading lamps with translucent paper shades in the fashionable Japanese style and a richly coloured rug beneath the billiard table adds warmth underfoot. A Baroque-style mirror with an ornately carved wooden frame is hung on the wall facing the door, between two substantial built-in bookcases housing part of Colonel Jenner's extensive library.

To find a selection of sporting trophies and stuffed animals, Russell visited the Newark Antique Market; he particularly admires the work of an early twentieth-century taxidermist called Rowland Ward, and some of his handiwork is evident in this room. In addition, there is a selection of military and sporting memorabilia, from pictures of Leopold Jenner on his pony to polo mallets and balls. The wind-up gramophone player would have provided musical accompaniment throughout the house, a revolutionary technological innovation which provided portable entertainment before the coming of radio.

RIGHT: The newly restored billiard table dominates a room that has been transformed into a masculine enclave.

BELOW: The room is filled with military and sporting paraphernalia to reflect Leopold Jenner's life.

A Zest for Life

ABOVE: Alexander Keiller concentrates over some plans in his Drawing Office at Avebury Manor.

Alexander Keiller used his considerable wealth to bankroll his archaeological investigations at Avebury, acquiring a total of 384ha (950 acres) of land in and around the village. The preservation of Avebury Manor and the stone circle today is largely due to Keiller's dedication to them over a number of years.

Dynamic and inspirational, technologically advanced and fond of fast cars and lively female company, Keiller sounds like a fictional hero; he was a champion ski-jumper, a pioneering aviator and an crack shot. He even tried his hand at being a detective, joining the Special Constabulary as an officer during the Second World War. He was also interested in witchcraft and the occult, and was married four times.

Keiller was the phenomenally wealthy heir to the highly successful Dundee-based business James Keiller & Sons, which was established in 1797 and made marmalade and other forms of preserves and confectionery, exporting their wares all over the world. Alexander's father had died when he was only nine and his mother when he was 17, which meant that he accessed his fortune at an early age. In 1913, at the age of 23, he married Florence, daughter of the artist Philip Richard Morris, and helped to found the Sizaire-Berwick motor manufacturing company, which produced luxury cars.

On the outbreak of the First World War Keiller joined the Royal Naval Volunteer Reserve, transferring to the Royal Naval Air Service in 1914. He was an early pioneer of flying and, despite being invalided out of the services in 1915, he joined air intelligence and stayed there until the end of the war. His passion for flying was later to combine with his interest in archaeology; in 1922, together with an archaeologist named O.G.S. Crawford, he conducted an aerial survey of sites in south-west England, and in 1928 they published *Wessex from the Air*, the first book of aerial archaeology to be published in Britain.

ABOVE: Keiller's millions came from the marmalade business established in Dundee at the end of the eighteenth century.

BELOW: Keiller flew a De Havilland DH.9, a First World War bomber, to carry out his aerial survey over south-west England.

Archaeological excavations at Avebury

Keiller's first marriage had ended in divorce after the war and in 1924 he married Veronica Mildred Liddell. Their shared interest in archaeology prompted them to visit Avebury later that same year, following which Keiller began to buy and excavate sites in and around the village, including Windmill Hill and West Kennet Avenue. He also acquired Avebury Manor in 1937, and much of the henge and the surrounding land. Every spring Keiller would return to Avebury to conduct further excavations, the findings of which were meticulously recorded. He began to find enormous standing stones which had been overturned and buried and, where possible, he re-erected the stones *in situ*; where they had disappeared completely he put a concrete marker to indicate where they had once stood.

Keiller's first major excavation of Avebury stone circle started in 1937, when he also moved the Morven Institute of Archaeological Research to Avebury Manor. His work continued for three years until the Second World War made further digs impossible. Each year he concentrated on a different quadrant of the site, locating, excavating and reinstating the massive sarsens of the stone circle. In 1938, his team discovered the skeleton of a man, who came to be known as the Barber-Surgeon on account of the scissors and medical probe found with him; judging by three coins among his remains he had been killed

by one of the stones in the early fourteenth century, possibly the 1320s. In 1938 Keiller also established the Alexander Keiller Museum, where artefacts found with the Barber-Surgeon are still on view.

Love and marriage

Alexander and Veronica divorced in 1934 and four years later Keiller married his third wife, an artist called Doris Emerson Chapman, who had joined the Institute. Doris was commissioned to paint an advert for Shell, the petrol company which took a pride in their avant-garde marketing techniques. She depicted one of Keiller's fellow diggers, holding the retrieved skull of the Barber-Surgeon. The Shell company decided that this was a somewhat gloomy image in the light of world events; the poster was not put into production, and the painting has since disappeared.

In June 1951, not long after an operation for throat cancer, Alexander Keiller divorced Doris and, the following day, married his fourth wife, Gabrielle Styles, the champion golfer and art collector. He died in 1955 at his home in Kingston Hill, Surrey, and in 1966 his widow donated the contents of the Alexander Keiller Museum to the nation. Keiller had been a pioneer in logging and interpreting historic sites – keeping detailed records of his finds, setting up a museum and putting up interpretation panels to engage the public – and made a lasting contribution to the archaeology of Avebury, which became a World Heritage Site in 1986.

THE KEILLER PARLOUR

This low-ceilinged, wood-panelled room on the east side of the manor sits in the oldest part of the house, which dates from *c.*1557, and probably once formed part of the single-storey Great Hall. The main entrance doorway to what was originally a fairly simple house, one room deep, still survives. The front door leads into a generously proportioned lobby, to the left of which is a fairly grand flight of stairs leading to the first floor. To the right the lobby opens up into what was traditionally known as the Little Parlour. The fireplace in the east wall is set at an oblique angle and provided heat for what was a possibly chilly east-facing room. The sash-windowed bay, with its deep window seat and view across the garden to the church, was added in the eighteenth century and retains much early crown glass.

BELOW: During the Jenners' ownership of Avebury Manor the Little Parlour was a cosy lobby filled with books.

In 1935 Alexander Keiller took out a lease on Avebury Manor, which he used as the base for his archaeological research, eventually buying the house in 1937. As his colleagues working on the excavations tended to be billeted at the nearby Red Lion Inn or Perry's Hotel, or in a variety of cottages nearby, Keiller often had guests for lunch or dinner, preceded by cocktails in either this parlour or the Tudor Parlour next door. His third wife, Doris, whom he married in 1938, was an accomplished modern artist and the couple divided their time between their London home, travels overseas and Avebury, which must have seemed an idyllic spot, isolated from a world hovering on the brink of war.

Alexander Keiller often had guests for lunch or dinner, preceded by cocktails in either this parlour or the Tudor Parlour next door.

Money was no object when it came to entertaining; it was remarked upon locally how many prehistoric objects intended for the museum were stored in wooden crates which had previously held bottles of Fortnum & Mason's best port. In more recent years, when a pond site in the manor grounds was cleared, it became apparent that it had been used as a rubbish dump for all sorts of inter-war packaging, most of it from culinary delicacies such as caviar. Predictably, the pond also contained many distinctive Keiller's Dundee Marmalade jars.

The parlour was used as a convenient open-plan reception area for visitors to the house, at all times of the day and night. Although austerely furnished in a sixteenth- and seventeenth-century style, it was also popular as a sitting room in the summer months, the key period for excavations.

BELOW: This curiously shaped room is built on two levels, with the fireplace set at an angle. The panelling was installed in the eary twentieth century by the Jenners.

TRANSFORMING THE
KEILLER PARLOUR

R ussell and the experts were keen to impart the contradictory nature of Alexander Keiller's confident, mercurial character; his love of the past and his immersion in the latest technology, his twin passions for ancient relics and high-speed engines, and his mission to catalogue minutely the mysterious artefacts of the remote past. This room, with its salvaged seventeenth-century wooden panelling (installed by the Jenners) and attractive bay window, had clearly been inhabited by a sophisticated, restless millionaire with modern tastes, and although Keiller himself chose austere period furniture, the team felt his interests and times were best reflected by transforming it in the Art Deco style.

The 1930s furnishings, with touches of bright colour, abstract patterns, and shiny surfaces, impose the Machine Age on a cosy, dark and venerable setting. Throughout, reference is made to Keiller's pastimes and interests – his pioneering aerial photography, his skiing, his involvement in motor manufacturing, and his fascination with natural forms such as shells and corals. As for his working relationships with his highly committed researchers, these could be represented by photographs.

RIGHT: In the 1970s, the parlour was furnished in a manner that reflected its early heritage, with Jacobean-style table and chairs.

242

Soft Furnishings

Ulster Carpet Mills were commissioned to make a bespoke carpet for this room, and it was a complicated brief. The firm's designer, Gemma Alexander, was supplied with visual references from Russell's studio, which included a design for a rug based on a late 1920s magazine graphic by American architect Frank Lloyd Wright and a silk headscarf with an abstract design of racing cars – the sort worn by a glamorous woman anxious to preserve her Marcel Wave while taking a spin in an open-top tourer. The carpet needed to encapsulate Keiller's passion for the modern age, so Gemma researched Rolls Royce cars of the 1930s and looked at avant-garde stained glass windows of the same era. The result is a futuristic interpretation of headlights and hubcaps, rendered as though seen through a prism or kaleidoscope.

Russell also supplied a palette of colours for this design, to which Gemma added some extra hues. The design is an all-over repeating abstract pattern, with each repeat occurring every 91cm (35in) horizontally and 113cm (44½in) vertically. The Axminster carpet was woven as a single piece on a loom, using a wool and nylon mixture of yarns with a high tuft density to make it resilient.

For the curtains and window cushions, Gainsborough Silk Weaving Company supplied 20m (22yd) of bespoke fabric, made of cotton and viscose in a damask weave. The 1930s-style design reveals Modernist and Cubist influences, and is a subtle and complex abstract design of rectilinear shapes with intersecting verticals and horizontals in varying shades of turquoise, blue, green, and silver.

ABOVE: The carpet combines elements from a rug based on a design by Frank Lloyd Wright and an Art Deco headscarf, resulting in an abstract pattern fitting for the Machine Age.

BELOW: The creels at the back of the looms at Ulster Carpet Mills create a colourful display.

ART DECO DESIGN

A rt Deco was a style that emerged between the two World Wars and was celebrated in the 1925 *Exposition Internationale des Arts Décoratifs* in Paris, which acted as a showcase for modern decorative arts. At the time it was known as Jazz Style and Jazz Moderne because it was linked with the era's music and nightlife; the term Art Deco, derived from the name of the Paris exhibition, was not coined until the 1960s. The new style bridged the gap between industry and the arts, encouraging artists, designers and craftspeople to combine functionality with excellence of manufacture in all aspects of material life, from paintings and sculpture to upholstery fabrics.

At first, Art Deco encompassed an eclectic, exuberant mix of cultural influences, incorporating Cubist paintings, Futurist graphics, African tribal art and Moorish architectural details;

ABOVE: A hand-decorated Meiping vase by Clarice Cliff shows the stylised patterns and 'jazz' colours that were typical of the Art Deco era.

BELOW: *Les Girls*, a *c.*1930 bronze and ivory sculpture by Romanian artist Dimitri Chiparus exemplifies an earlier, more elite Art Deco style.

stylized motifs of fruits, flowers, dancing figures, wild animals and birds were combined with abstract geometric and architectural forms. In Britain, Ancient Egyptian art and architecture influenced Art Deco, following the archaeologist Howard Carter's discovery of the tomb of Tutankhamun in 1922. The Egyptian style suited architects seeking a new idiom for buildings with few precursors, such as cinemas.

The brilliant colours of the Ballets Russes changed the fashionable palette of the 1920s. Cream and ecru were offset by sharp orange, lacquer red, jade green, turquoise and black. Clarice Cliff, the British ceramic designer whose work remains much in demand today, developed avant-garde geometric forms for tea sets and vases and had them hand-decorated in 'jazz' colours with stylised rustic scenes, linked by sweeping parallel curves and abstract motifs.

At its most extreme, American Art Deco offered a Utopian vision of the future, with cars, planes, and fantasy buildings. Everyday objects such as vacuum cleaners and Bakelite radios celebrated the sleek new materials, metallic glossy surfaces and streamlined forms of the modern age, even applied to humdrum objects such as pencil sharpeners and coal scuttles.

By the 1930s, the emphasis was on functionality and efficiency; simpler, more elegant forms reflected more cerebral aspirations and the idea of the house being 'a machine for living'. The colour range in this decade was richer and more subdued and materials such as brass, glass and chrome were used in combination with decorative wood effects and varied finishes such as lacquer for interiors. Furniture became more compact, multi-functional and streamlined, reflecting the aesthetic appeal of ocean-going liners and their atmosphere of personal comfort and suave luxury.

BELOW: Art Deco influenced the way people decorated their homes. This bathroom at Upton House, Warwickshire, has aluminium leaf walls, black lacquered skirting boards and red lacquered pillars.

Painting Inspiration

For the alcove, artist Corin Sands painted an interpretation of Paul Nash's colour lithograph of *Landscape with Megaliths*, dating from 1937. Nash, a supremely accomplished artist, first visited Avebury in 1933 and described the standing stones as 'wonderful and disquieting'. He was to produce at least three major artworks based on his visits to the stones; he was endlessly fascinated by the mystical qualities of inanimate objects, particularly when they dominated their surroundings. Coincidentally, Nash painted another prehistoric site, Wittenham Clumps, which had belonged to William Dunch, owner of Avebury Manor in the sixteenth century.

The Art Deco Screen

BELOW AND RIGHT: Specialist painter Mark Sands paints an Art-Deco-style, four-panel screen for the Keiller Parlour.

Corin's brother, specialist painter Mark Sands, was also commissioned by Russell to create a spectacular piece for this room – a four-panel folding screen, decorated with an exuberant Art Deco design. For inspiration,

Mark referred to a cast aluminium panel, now in the Victoria and Albert Museum, which had been part of a decorative frieze over a lift entrance in the Derry & Toms building in Kensington High Street, London. The original piece was probably designed by Walter Gilbert for the iconic department store, which loomed like a giant ocean-going liner over its more reticent neighbours in the 1930s. In more recent years, the former Derry & Toms was the home of Biba, the apotheosis of the 1970s Art Deco Revival.

Mark designed a decorative scheme to represent fertility, growth and the natural world, apparently springing from a stylised waterfall which arcs upwards from the centre of the two middle panels. Drawing on previous restoration work he had done on various Art Deco buildings in Miami, Mark agreed with Russell a range of authentically 1930s colours he felt could be added to the more limited palette range originally

ABOVE: Various pieces were bought at auction to add the finishing touches to the Keiller Parlour, such as this Art Deco statuette of an exotic dancer.

envisaged for the screen. His touches of emerald and lilac brought the design alive, and he provided paint samples to Ulster Carpet Mills to help inform their design for the carpet.

The screen was deceptively simple in its construction, with four rectangular pine frames, each filled with a panel of hardboard, hinged, then covered with calico and treated to take paint. Earthborn Paints supplied the paints, and specialist glazes were applied to the final surface to subdue the colours and make the screen look a little older. To add to the resilience of the finish, Mark applied a compound rather like linseed oil, but a quick-drying resin substitute. The decorative side of the screen faces the Parlour; the reverse, painted black with a scumbled effect of brown, faces the wood-lined service corridor running down to the kitchen, affording the guests some privacy from the passage of servants between the dining room and the kitchen. On this wall, which might otherwise be dimly lit, are some impressive Art Deco wall-lights purchased by Russell at the Newark Antiques Fair.

The Finishing Touches

In the alcove is a 1930s combined sofa and book cupboard, a versatile piece of furniture of the type much admired in Continental Europe, and originally associated with the Bauhaus, the radical art and design school established in pre-Nazi Germany. A pair of low armchairs in what was known as the 'liner style', with exaggerated curves and upholstered in a figured moquette in autumnal colours, flank a coffee table, completing the seating area on this side of the room. There is much emphasis on electric lighting, with a pair of floor lamps, a standard lamp, a table lamp and a window lamp to light this rather dark room to modern standards. Alexander Keiller's predecessors would undoubtedly have struggled to read in this room on dull winter days.

In the wide vestibule is a low 1930s armchair with wide wooden arms, which served as a useful spot to deposit packages, coats and handbags when arriving or departing. Nearby is that other essential to 1930s hospitality – a well-stocked cocktail cabinet, its curved frontage and smooth surfaces veneered in satinwood, complete with a cocktail shaker and a wide variety of glassware.

ABOVE: An original 'liner style' armchair with exaggerated curves.

OVERLEAF: The Keiller Parlour is designed for entertaining.

KEILLER ENTERTAINS

Keiller always maintained a spirit of hospitality and welcomed many guests to Avebury Manor. Meals were ample and delicious, and usually preceded by wines and cocktails served by the under-butler. He frequently entertained members of his excavation team and his museum staff, and one time deeply embarrassed his cook by requesting dinner be served at 8pm; the cook was completely unprepared, being more accustomed to serving the evening meal as late as 11pm to accommodate the lengthy 'cocktail hour'. Keiller also welcomed many strangers to the Manor including, on the first day of the Second World War, more than 70 children and five of their teachers who had arrived at Avebury after being evacuated from the East End of London.

Useful Addresses

Avebury

Avebury Manor House and Garden, Avebury Stone Circle, and Alexander Keiller Museum
National Trust Estate Office
High Street, Avebury,
Wiltshire SN8 1RF
www.nationaltrust.org.uk/avebury
01672 539250

Heritage Organizations

The National Trust
PO Box 39,
Warrington, WA5 7WD
www.nationaltrust.org.uk
✉ enquiries@nationaltrust.org.uk
0844 800 1895

English Heritage
The Engine House,
Fire Fly Avenue,
Swindon SN2 2EH
www.english-heritage.org.uk
✉ customers@english-heritage.org.uk
0870 333 1181

SPAB (The Society for the Protection of Ancient Buildings)
37 Spital Square, London E1 6DY
www.spab.org.uk
✉ info@spab.org.uk
tel 020 7377 1644

HHA (Historic Houses Association)
2 Chester Street,
London SW1X 7BB
www.hha.org.uk
✉ info@hha.org.uk
020 7259 5688

Craftspeople, Artists and Makers

involved in the transformation of Avebury Manor

Angus Handloom Weavers
The last handloom linen weaver working in Britain
Contact: Ian Dale, master weaver
House Of Dun, Montrose,
✉ angusweaversltd@aol.com
01674 810 255

Guy Butcher
Furniture maker and designer
Hansnett Farm, Canon Frome,
Ledbury, Herefordshire HR8 2TF
www.guybutcherfurniture.co.uk
✉ info@guybutcherfurniture.co.uk
01531 670160

Earthborn Paints
Specialist paints
Contact: John Dison
Frodsham Business Centre,
Bridge Lane, Frodsham,
Cheshire WA6 7FZ
www.earthbornpaints.co.uk
✉ info@earthbornpaints.co.uk
01928 734171

ELG at Sainsburys Ltd
Contact: Jonathan Sainsbury
Bespoke manufacturers
of fine carved & gilded
mirrors & furnishings
26 Old Street,
Bailey Gate, Sturminster Marshall,
Dorset BH21 4DB
www.elgsainsburys.com
✉ info@elgsainsburys.com
01258 857 573

Farrow & Ball
Specialist paints
www.farrow-ball.com

The Four Poster Bed Company
Makers of handmade four-poster beds and other furniture
Contact: Stephen Edwards
New House Farm, Lyonshall, Kington,
Herefordshire HR5 3JS
www.fourposterbed.co.uk
✉ stephen@fourposterbed.co.uk
01544 340 444

Fromental
Makers of handmade wallpapers and fabrics
326 Kensal Road, London W10 5BZ
(by appointment only)
www.fromental.co.uk
✉ info@fromental.co.uk
020 3410 2000

The Gainsborough Silk Weaving Company Ltd
Specialists in historical replicas
Alexandra Road, Sudbury,
Suffolk CO1O 2XH
www.gainsborough.co.uk
✉ sales@gainsborough.co.uk
01787 372081

Neville Griffiths
Specialist in architectural salvage
Rococo, 4–6 Church Street,
Lower Weedon,
Northampton NN7 4PL
www.nevillegriffiths.co.uk
✉ nevillegriffiths@mac.com
07872 822610

Henry Newbery
Creators and suppliers of fine furnishing trimmings and fabrics
Unit 7G, Regent Studios,
1 Thane Villas, London N7 7PH
www.henrynewbery.com
✉ sales@henrynewbery.com
020 7281 5088

Hubble Sports
Snooker and billiard table suppliers and restorers
Contact: Peter Ludgate
Fairbourne Court, Harrietsham,
Kent ME17 1LQ
www.hubblesports.com
✉ hubble.sports@gmail.com
01622 859776

Hughes Woodcarving
Contact: Emyr Hughes,
master carver
Unit 12, Staunton Court,
Staunton, Gloucestershire
GL19 3QE
www.hugheswoodcarving.co.uk
01452 840 803

Keramis Lighting and Porcelain
Makers of bespoke Chinese porcelain
Contact: Benjamin Creutzfeldt
www.keramis.net
✉ benjamin@creutzfeldt.net

Gudrun Leitz
Green woodwork tutor and furniture maker
Hill Farm, Stanley Hill, Bosbury,
Ledbury, Herefordshire HR8 1HE
www.greenwoodwork.co.uk
✉ gudrun@greenwoodwork.co.uk
01531 640125

Dave Lyons
Specialist furniture restorer
07790 234849

Royal School of Needlework
Apartment 12a, Hampton Court
Palace, Surrey KT8 9AU
www.royal-needlework.org.uk
✉ enquiries@royal-needlework.org.uk
020 3166 6932

Rushmatters
Makers of traditional
handcrafted rush flooring and other rush products
Contact: Felicity Irons
Grange Farm, Colesden, Bedfordshire
MK44 3DB
✉ felicityirons@rushmatters.co.uk
01234 376419

Corin Sands
Fine artist and restorer
www.corinsands-art.de
✉ corinsands@hotmail.com
+49 (0)6581 998494

Mark Sands
Specialist painter and restorer
www.markbensonsands.com

George Smith Ltd
Manufacturers of handmade sofas
587–589 King's Road,
London SW6 2EH
www.georgesmith.co.uk
✉ sales@georgesmith.co.uk
020 7384 1004

Thomasina Smith
Artist and illustrator
✉ thom@parkstudios.demon.co.uk
07967 984 593

Ulster Carpets
Carpet manufacturers
Castleisland Factory, Craigavon BT62
1EE,Northern Ireland
www.ulstercarpets.com
✉ marketing@ulstercarpets.com
028 3833 4433

Grant Watt
Fine decorator
✉ grant_scenic@yahoo.co.uk
07739 794211

Zardi & Zardi Ltd
Makers of copies of tapestries and historic textiles
Contact: PJ Keeling
Podgwell Barn, Sevenleaze Lane, Edge,
Stroud, Gloucestershire
GL6 6NJ
http://www.zardiandzardi.co.uk
✉ enquiries@zardiandzardi.co.uk
01452 814 777

INDEX

Acknowledgements

A large number of people were helpful in providing information and expertise while this book was in preparation. The author would like to thank the following people who were especially generous with their time and knowledge: Lucy Armstrong, Michelle Bullen, Guy Butcher, Tim Butcher, Ros Cleal, Benjamin Creutzfeldt, Dan Cruickshank, Alison Dalby, Ivan Day, James Dobson, Neville Griffiths, Siobhan Griffiths, Charlotte Gittins, John Hammond, Andre Holzinger, Chris Lacey, Jill Lovett, Peter Ludgate, Jilly MacLeod, Thomas Sainsbury, Diane Sargeant, Sarah Staniforth, Rodney Timpson, Jan Tomlin, Grant Watt, Thomas Woodcock.

Picture Credits

Anova Books is committed to respecting the intellectual property rights of others. We have therefore taken all reasonable efforts to ensure that the reproduction of all content on these pages is done with the full consent of copyright owners. If you are aware of any unintentional omissions, please contact the company directly so that any necessary corrections may be made for future editions.

Front cover: BBC/ Colin Bell
Back cover top: National Trust Images/ Paul Wakefield
Back cover bottom: National Trust/ James Dobson

Page 45 ©Alamy/David Lyons; Page 49 top ©Alamy/ The Art Gallery Collection; Page 63 ©Alamy/ Classic Image; Page 65 top ©Alamy/Ed Pavelin; Page 101 bottom Alamy/ Robert Harding Picture Library Ltd; Page 116 bottom ©Alamy/ Guy Edwardes Photography; Page 146 bottom ©Alamy/ North Wind Picture Archives; Page 169 top ©Alamy/ Lebrecht Music and Arts Photo Library; Page 213 ©Alamy/Pictorial Press Ltd

Pages 32 left, 135, 190 top, 221 top, 236, 238, 239 © Alexander Keiller Museum

Page 92 ©Angus Handloom Weavers

Page 131 bottom ©Anova Books

Pages 02–03, 21, 124, 125, 136 top, 137, 195, 200, 203 bottom, 230, 246, 247, 248, 249 top ©BBC/Andre Holzinger; Page 17 © BBC/Colin Bell; Pages 76 bottom, 84 bottom, 94, 123 bottom, 196 top, 204, 205 ©BBC/ Thea Hey; Page 77 bottom, 147, 151 bottom ©BBC/Sophie George; Page 85 © BBC/ Charlotte Gittins; Pages 91, 188, 189 top ©BBC/ Jon Eastman

Page 29 top ©The Bridgeman Art Library/Ashmolean Museum, University of Oxford, UK; Pages 44 top, 175 top ©The Bridgeman Art Library/ Private Collection ; Pages 98 bottom, 108 top ©The Bridgeman Art Library/ The Crown Estate; Page 101 top ©The Bridgeman Art Library/ National Portrait Gallery, London, UK; Page 148 ©The Bridgeman Art Library/ Private Collection/ Archives Charmet; Page 170 bottom ©The Bridgeman Art Library/ The Museum of London; Page 171 ©The Bridgeman Art Library/ Guildhall Library, City of London; Pages 178, 244 top ©The Bridgeman Art Library/ Private Collection/ Christopher Wood Gallery, London, UK; Page 211 ©The Bridgeman Art Library/Bibliotheque des Arts Decoratifs, Paris, France/ Archives Charmet; Page 214 bottom ©The Bridgeman Art Library/The Advertising Archives; Page 215 ©The Bridgeman Art Library/ Private Collection/The Stapleton Collection; Page 244 bottom © The Bridgeman Art Library /Private Collection/ © Dreweatt Neate Fine Art Auctioneers, Newbury, Berks, UK

Page 79 ©Cadworks UK Ltd

Page 64 ©Charney Manor. Photograph taken by Sylvia Gibbons

Page 154 top ©Coutts & Co. Photograph taken by BBC/ Olly Gordon.

Pages 66, 84 top, 118, 150, 219 bottom, 226, 228, 240 © Country Life Picture Library

Page 130, 143 bottom ©ELG at Sainsburys Ltd; Page 158 top ©ELG at Sainsburys Ltd. Photograph taken by BBC/Thea Hey

Pages 116 top, 146 top © Fitzwilliam Museum, Cambridge

Pages 154 bottom, 155, 157 ©Fromental. Photographs taken by BBC/ Thea Hey

Pages 128 top, 128 bottom, 129, 144 © Gainsborough Silk Weaving Company. Pages 8–9 © Gainsborough Silk Weaving Company. Photograph taken by National Trust/James Dobson

Pages 73, 74 top, 74 bottom, 75 left, 75 right, 76 top ©Guy Butcher

Page 237 bottom ©Imperial War Museum, London

Page 237 top ©Jilly MacLeod

159 bottom ©Keramis Lighting and Porcelain

Pages 34–35 ©Martin Brown

Pages 30, 44 bottom, 185 bottom, 187 top, 214 top, 233 bottom © Mary Evans Picture Library; Page 32 right © MEPL/ Francis Frith; Pages 208–209, 212 top ©MEPL/ Retrograph Collection; Page 210 ©Mary Evans Picture Library/ONSLOW AUCTIONS LIMITED; Page 222 ©Mary Evans Picture Library/ Illustrated London News Ltd

Page 221 bottom ©The Museum of Modern Art, New York/Scala, Florence

Pages 4, 11, 12, 24–25, 26, 28, 33, 35 top, 36 bottom, 37 top, 37 bottom, 39 top, 39 bottom, 40, 41 top, 41 bottom, 46–47, 62, 65 bottom, 82 top, 82 bottom, 83, 89 top, 119 top, 121, 122 bottom, 140 top, 142, 189 bottom, 191, 194 top, 196 bottom, 198 top, 198 bottom, 199, 203 top, 206–207, 219 top, 223 bottom, 227, 231 bottom ©National Trust/ James Dobson; Pages 6, 7, 14, 18, 19, 23, 67 bottom © NT/Chris Lacey; Page 31 left © NT/ Colin Brown; Pages 80–81, 95, 132–133, 145, 164–165, 192–193, 234, 235 ©NT/ John Hammond

Page 224 ©The National Trust at Lytes Cary Manor

Page 115 bottom, 161 bottom, 166–167, 223 top © National Trust Images; Page 10 ©NTI/ Flo Smith; Pages 15, 29 bottom, 51, 57, 60 top, 86 bottom, 88, 90, 106, 112 top, 115 top, 127 top, 127 bottom, 134, 138, 139 top, 139 bottom, 140 bottom, 153, 160, 173, 174, 176 top, 179, 180 top, 180 bottom, 181, 184, 186, 187 bottom, 194 bottom, 201, 220, 232, 241, 245, 249 top ©NTI/ Andreas von Einsiedel; Page 129 © NTI/Andreas von Einsiedel/National Museums Liverpool (Lady Lever Art Gallery); Pages 20, 229 © NTI/ James Dobson; Page 27, 152 top ©NTI/ David Levenson
Pages 38 top, 49 bottom, 56 bottom, 58, 59 bottom, 71, 100, 105, 107 bottom, 112 bottom, 123 top, 141 top, 152 bottom, 161 bottom, 169 bottom, 185 bottom background, 211 bottom ©NTI/ John Hammond; Page 38 bottom ©NTI/ Stephen Robson; Page 42 ©NTI/ David Noton; Page 43 ©NTI/ John Heseltine; Pages 48, 107 top, 168 ©NTI/ Derrick E Witty; Page 50 top ©NTI/ Charlie Waite; Pages 50 bottom, 70 top, 122 top ©NTI/ Angelo Hornak; Pages 52, 53 bottom, 56 top ©NTI/ Robert Morris; Pages 53 top, 55, 70 bottom, 87, 89 bottom, 109, 113, 170 top, 172, 175 bottom, 177, 218 ©NTI/ Nadia Mackenzie; Page 54, 60 bottom, 86 top © NTI/ Nick Guttridge; Page 59 top ©NTI/ Brenda Norrish; Page 61 ©NTI/Cristian Barnett; Page 68 top ©NTI/ John Bethell; Page 69 ©NTI/ Geoffrey Fosh; Page 71 ©NTI/ Andrew Haslam; Pages 96–97, 98 top, 103 top, 111 bottom ©NTI/ Bill Batten; Page 99, 217 ©NTI/Nick

Meers; Page 103 bottom ©NTI/ Christopher Hurst; Page 104 ©NTI/Rupert Truman; Page 108 bottom ©NTI/ David Hall; Page 110 ©NTI/ David Miller; Page 111 top ©NTI/ Matthew Antrobus; Page 114 ©NTI/ James Mortimer; Page 119 bottom, 126, 216 top, 216 bottom, 231 top, 233 top, 249 bottom ©NTI/ Dennis Gilbert; Pages 143 top, 151 top ©NTI/ Mark Fiennes; Page 176 ©NTI/ Rob Talbot; Page 197 top ©NTI/ Jonathan Buckley; Page 197 bottom ©NTI/ Paul Harris; Page 202 ©NTI/NaturePL/David Kjaer

Pages 13, 117 ©National Trust Magazine/Cristian Barnett

Page 163 bottom, 190 bottom ©Rodney Timpson

Page 78, 131 top ©The Royal Collection/ 2011 Her Majesty Queen Elizabeth II

Page 77 top ©Rushmatters

Pages 120, 141 bottom, 242 © Savernake Estate

Page 68 bottom ©Selly Manor and Bournville Village Trust

Pages 29 top background, 37 bottom background ©Shutterstock Stock Images

Page 149 ©The Trustees of the British Museum

Pages 162, 163 top, 243 top, 243 bottom ©Ulster Carpets

Page 136 bottom ©V&A Images/ Victoria and Albert Museum

Page 36 top ©Wiltshire and Swindon Archives

Pages 67 top, 182, 183 top, 185 top, 200, 212 bottom ©Wiltshire Heritage Museum Archive & Library, Devizes. Photographs taken by BBC/Andre Holzinger; Page 183 ©Wiltshire Heritage Museum Archive & Library, Devizes. Photograph taken by BBC/Sophie George